Automotive Engines: An Engineering Perspective

Automotive Engines: An Engineering Perspective

Alice Martin

MURPHY & MOORE
www.murphy-moorepublishing.com

Published by Murphy & Moore Publishing,
1 Rockefeller Plaza,
New York City, NY 10020, USA

ISBN: 978-1-63987-068-4

Cataloging-in-Publication Data

Automotive engines : an engineering perspective / Alice Martin.
p. cm.
Includes bibliographical references and index.
ISBN 978-1-63987-068-4
1. Automobiles--Motors. 2. Automobiles--Motors--Maintenance and repair.
3. Automobiles--Design and construction. I. Martin, Alice.
TL210 .A98 2022
629.250 4--dc23

For information on all Murphy & Moore Publications
visit our website at www.murphy-moorepublishing.com

MURPHY & MOORE

Contents

Preface VII

Chapter 1 **Automotive Engines: An Introduction** 1

Chapter 2 **Internal and External Combustion Engine** 27
i. Internal Combustion Engine 27
ii. External Combustion Engine 77

Chapter 3 **Boost Controller and Engine Sensors** 79
i. Boost Controller 79
ii. Engine Sensors 81

Chapter 4 **Mechanics of Combustion** 89
i. Combustion 89
ii. Combustion Triangle 101
iii. Fuel 104
iv. Thermodynamics of Combustion 118
v. Thermodynamics 128
vi. Stoichiometry 146
vii. Conservation of Mass 158

Chapter 5 **Significant Aspects of Automotive Engines** 167
i. Vehicle Emissions Control 167
ii. Automotive Electronics 171
iii. Automatic Transmission 174
iv. Manufacturing Engineering 191
v. g-force 202
vi. Fuel Efficiency 211
vii. Safety Engineering 218
viii. Engine Power 222
ix. Revolutions Per Minute 228

Permissions

Index

Preface

An engine is a device which can be used to convert one form of energy into mechanical energy. Automotive engines are those which can be used for propelling an automobile. According to the energy conversion method, they are broadly classified into heat engines, electric motors and physically powered motors. Heat engines are thermodynamic systems which can convert thermal energy of fuel into mechanical energy. They are further categorized into internal combustion engine, external combustion engine and air breathing combustion engine. Ramjet, pulse jet, scramjet, gas turbines, steam turbines and reciprocating engines are some of its prominent examples. This book unfolds the innovative aspects of automotive engines which will be crucial for the holistic understanding of the subject matter. The topics included herein on this subject are of utmost importance. Those in search of information to further their knowledge will be greatly assisted by this book.

A detailed account of the significant topics covered in this book is provided below:

Chapter 1- Automotive engines are a wide-variety of heat and electric engines which are used in automotive vehicles. Some of the important topics studied in automotive engines are fuel supply systems, catalytic converters and charging of engines. This is an introductory chapter which will introduce briefly all these significant aspects of automotive engines.

Chapter 2- There are two main types of heat engines: internal combustion engine and external combustion engines. Internal combustion engines are further divided according to their design, fuel used, cycle of operation, number of strokes, ignition, etc. while external combustion engines are divided into steam and stirling engines. All these diverse types of engines have been carefully analyzed in this chapter.

Chapter 3- A device that is used to regulate the boost level produced by a supercharged engine by regulating the air pressure carried out to the pneumatic and mechanical wastegate actuator is known as boost controller. The topics elaborated in this chapter will help in gaining a better perspective about boost controller as well as different types of engine sensors such as engine coolant temperature sensor, engine speed sensor (ESS) and oxygen sensor.

Chapter 4- Combustion refers to the exothermic redox chemical reaction that takes place between a fuel and an oxidant giving off heat and light in the process. Some concepts related to combustion such as combustion triangle, fuel, thermodynamics of combustion, stoichiometry and conservation of mass are discussed in this chapter in an easy-to-understand manner.

Chapter 5- Some of the important aspects of automotive engines are vehicles emissions control, automotive electronics, automatic transmission, manufacturing engineering, g-force, fuel efficiency, safety engineering, engine power and revolutions per minute. This chapter has been carefully written to provide an easy understanding of these facets of automotive engines.

It gives me an immense pleasure to thank our entire team for their efforts. Finally in the end, I would like to thank my family and colleagues who have been a great source of inspiration and support.

Alice Martin

1

Automotive Engines: An Introduction

Automotive engines are a wide-variety of heat and electric engines which are used in automotive vehicles. Some of the important topics studied in automotive engines are fuel supply systems, catalytic converters and charging of engines. This is an introductory chapter which will introduce briefly all these significant aspects of automotive engines

Automobile is a vehicle driven by an internal combustion engine and it is used for transportation of passengers and goods on the ground. Automobile can also be defined as a vehicle which can move by itself. Examples: Car, jeep, bus, truck, scooter, etc.

Classification of Vehicles

Automobiles or vehicles can be classified on different bases as given below:

- On the basis of load:
  - Heavy transport vehicle (HTV) or heavy motor vehicle (HMV), e.g. trucks, buses, etc.
  - Light transport vehicle (LTV), e.g. pickup, station wagon, etc.
  - Light motor vehicle (LMV), e.g. cars, jeeps, etc.
- Wheels:
  - Two wheeler vehicle, for example: Scooter, motorcycle, scooty, etc.
  - Three wheeler vehicle, for example: Auto rickshaw, three wheeler scooter for handicaps and tempo, etc.
  - Four wheeler vehicle, for example: Car, jeep, trucks, buses, etc.
  - Six wheeler vehicle, for example: Big trucks with two gear axles each have four wheels.
- Fuel used:
  - Petrol vehicle, e.g. motorcycle, scooter, cars, etc.
  - Diesel vehicle, e.g. trucks, buses, etc.
  - Electric vehicle which use battery to drive.
  - Steam vehicle, e.g. an engine which uses steam engine. These engines are now obsolete.
  - Gas vehicle, e.g. LPG and CNG vehicles, where LPG is liquefied petroleum gas and CNG is compressed natural gas.

- Body:
  - Sedan with two doors.
  - Sedan with four doors.
  - Station wagon.
  - Convertible, e.g. jeep, etc.
  - Van.
  - Special purpose vehicle, e.g. ambulance, milk van, etc.
- Transmission:
  - Conventional vehicles with manual transmission, e.g. car with 5 gears.
  - Semi-automatic.
  - Automatic: In automatic transmission, gears are not required to be changed manually. It is automatically changes as per speed of the automobile.
- Position of Engine:
  - Engine in Front: Most of the vehicles have engine in the front. Example: most of the cars, buses, trucks in USA.
  - Engine in the Rear Side: Very few vehicles have engine located in the rear.

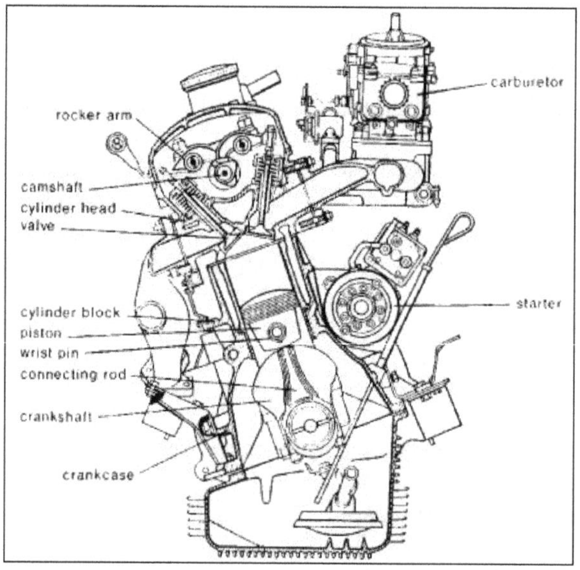

Transverse cross section through the MZMA-412 carburetor engine.

Automotive engines consist of multiple cylinders (usually four, six or eight), all bored into a block of metal (the engine block). Each cylinder contains a piston, which moves up and down. A connecting rod connects the piston to a rotating crankshaft, which turns the up-and-down motion of the piston into a circular motion. Automotive engines are divided into four-cycle and two-cycle engines according to the working process and are divided by the method of fuel ignition into spark plug ignition engines (also known as carburetor engines or gasoline engines) and diesel engines, which are self-ignition engines using high-temperature air compressed within the engine cylinders. The fuel mixture of vaporized gasoline and air, blended in the carburetor, is admitted to the cylinders of carburetor piston engines. Piston engines without carburetors also exist; these are equipped with a device for direct injection of fuel into the intake manifold or into the engine cylinder.

The operating cycle of these engines is the same as that of carburetor engines. In diesel engines, diesel fuel is sprayed through the nozzle of a high-pressure pump directly into the cylinders, where it is mixed with air. Automotive engines are distinguished by the number and placement of the cylinders (in-line engines, V-engines, and so on), by the location of the valves (in the cylinder head or in the cylinder block), by the cylinder capacity, by the motor cooling system (liquid or air), by function, and so on. Four-cycle overhead valve liquid-cooled carburetor-type piston engines are used in modern automobiles, as well as in small and medium trucks. Diesels, which burn a cheaper fuel than gasoline and offer advantages over carburetor type engines in fuel economy and length of service, are used mainly for the propulsion of heavy trucks and large buses. Gasoline motors are superior in ease of design and low initial cost, power per liter of displacement, starting qualities, and control of exhaust smoke, but modern high-speed diesel engines come close to carburetor engines in such important performance characteristics as specific mass (kg/kW or kg/hp), compactness, and noise-free operation. With these advances, and thanks to increases in power per liter of displacement, diesel engines have begun to find applications within the past decade in lightweight trucks and even in passenger automobiles.

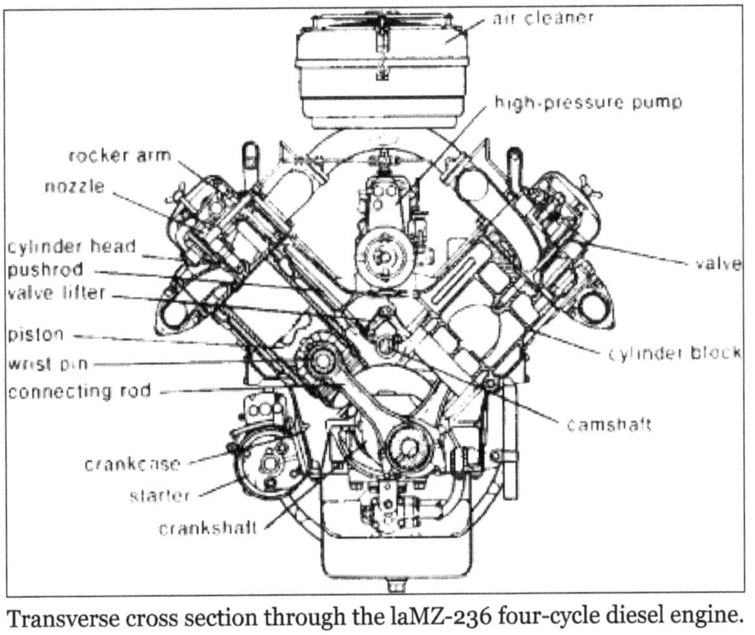

Transverse cross section through the laMZ-236 four-cycle diesel engine.

Modern four-cycle piston engines consist of a cylinder block usually made integrally with the crankcase, the cylinder head, pistons with sealing rings and oil-control piston rings, connecting rod, crankshaft, flywheel, camshaft, spring-loaded intake and exhaust valves, valve train parts (rocker arm, valve lifter), the drive connecting the crankshaft and camshaft, spark plugs or fuel nozzle, and so on. These engines are equipped with a radiator and fan for the cooling system. They have pumps for forced circulation of lubricant oil and cooling fluid and for bringing fuel from the fuel tank, as well as fuel filters, oil filters, and air cleaners, a starter or cranking motor, ducts for air, gas, fuel, oil, and cooling fluid, and automatic control devices controlling the frequency of rotation of the crankshaft and the temperature of the cooling fluid and fuel mixture.

The engine most essential part of automobile industries or we can say that the engine is the heart of an automobile. The function and construction of each engine parts of an internal combustion engine are explained. The key to the engine is as follows. The main car engine parts are as follows:

- Cylinder Block
- Cylinder Head
- Crank Case
- Oil Pan
- Manifolds
- Gasket
- Cylinder Liner
- Piston
- A Piston Ring
- Connecting Rod
- Piston Pin
- Crank Shaft
- CamShaft
- Flywheels
- Engine Valves:
    - Poppet valve
    - Sleeve valve
    - Rotary valve
- Governers

Cylinder Block

Figure shows a simple sketch of the cylinder block. It is the basic framework for the engine. and one of the main in engine parts. Cylinder block, cylinder head and crankcase these three parts form the foundation and main stationary body of the automobile engine.

- A cylinder block consists of three parts:
    - The cylinder in which the piston slid up and down.
    - The port or opening for the valves.
    - The passages for the flow of cooling water.

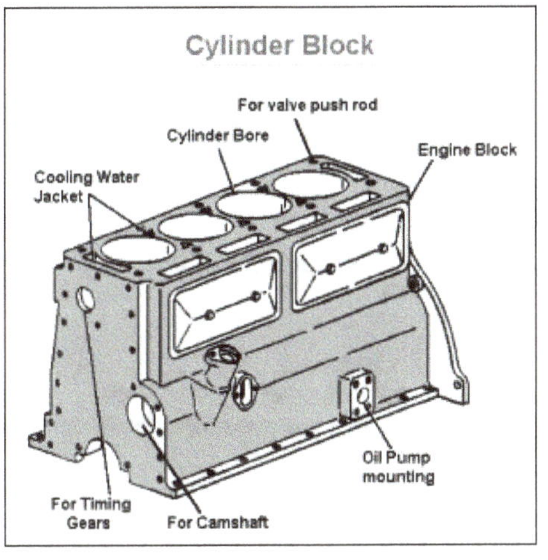

- Construction and Working:
    - The cylinder block is usually made of grey cast iron or aluminium and its alloys.
    - While the crankcase is fixed to its bottom. Apart from these other parts like timing gear water pump, ignition distributor, flywheel, fuel pump etc., are also attached to it.
    - Passages are provided in the cylinder walls for the circulation of cooling water.
    - Mating surfaces of the block are carefully machined to provide a perfect sealing surface.
    - Cylinder block also carries lubrication oil to various components through drilled passages called oil galleries.

Cylinder Head

Another type of engine parts is cylinder head it is joint between the cylinder head and cylinder block.

- Construction:

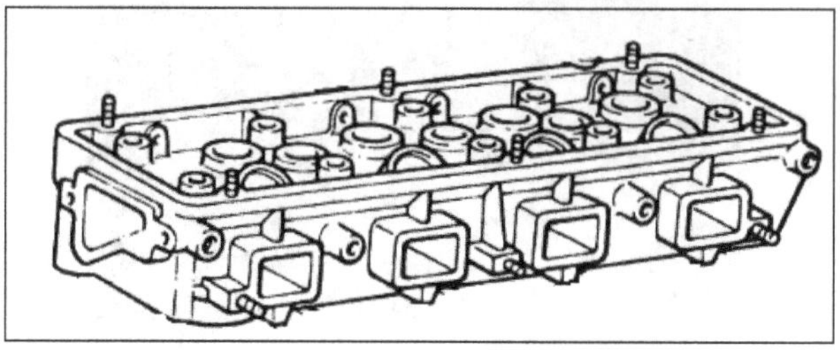

Cylinder head

  - It is usually made up of cast iron and aluminium alloy.
  - The top of the cylinder is covered by a separate cast piece known as the cylinder head.
  - The cylinder head is attached to the cylinder block by means of studs fixed to the block gaskets are used to provide a tight, leak-proof joint between the head and block.
  - Cylinder head contains a combustion chamber above each cylinder.
  - It also contains valve guides, valve seats, ports, coolant jackets and threaded holes for spark plugs. It incorporates passages for the flow of cooling water.

- Applications:

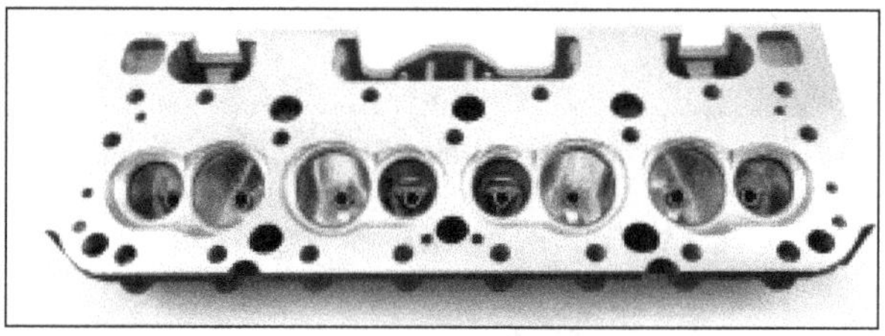

  - The cylinder head casts integrally with cylinder blocks also be done in a few cases usually in racing cars to obtain a gas-tight joint.
  - The detachable head types are more advantages than integral construction.
  - However, for certain heavy-duty engine requires high cooling rates such as in racing cars copper alloys may be used.

- Types of Cylinder Head: Depending upon the valve and port layout, the cylinder head may be classified into three types as follows:
  - Loop flow type: In the loop, flow types the inlet and the exhaust manifolds are on the same side, which facilitates preheating of the intake air.
  - Offset cross flow type: Offset cross flow types the inlet and the exhaust manifolds are placed on different sides of the cylinder head.
  - In line cross flow type: In line cross-flow type, the valve is positioned transversely and usually inclined to each other, while the inlet and the exhaust manifolds are on different sides of the cylinder head. This arrangement gives better performance, but it is costlier.

Crankcase

The oil pan and the lower part of the cylinder block together are called the crankcase. It is the bottom portion of the cylinder block, in which the crankshaft is fitted.

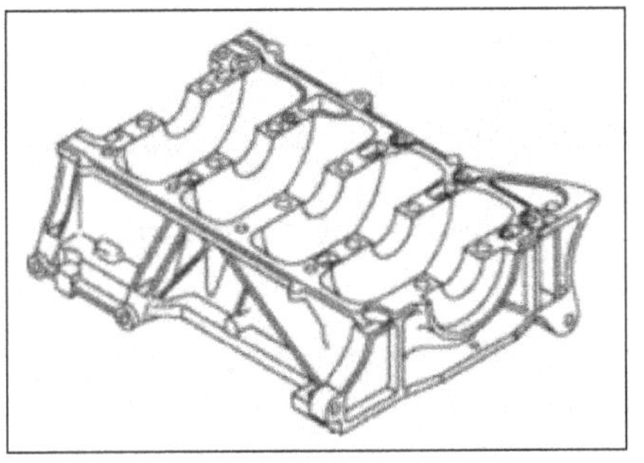

Crankcase

- Construction:
  - This is a rigid construction made of grey cast iron or aluminium. Either it can be cast integrally with the block or can be cast separately and attached to the block with bolts.
  - The crankcase is shaped simply like a box having no bottom. Oil pan or sump forms the bottom half of the crankcase.
- Working:
  - The function of the crankcase is to provide support for the main journals and bearing of the crankshaft, rigidly maintaining the alignment of their axes of rotation under various engine loads.
  - The crankcase is supported in the crankcase through a number of bearing called the main bearing.

Oil Pan

The bottom half of the crankcase is called the oil pan or sump. It is attached to the crankcase through set screws and with a gasket to make the joint leak proof. The oil pan serves as a reservoir for the storage, cooling and ventilation of engine lubricating oil.

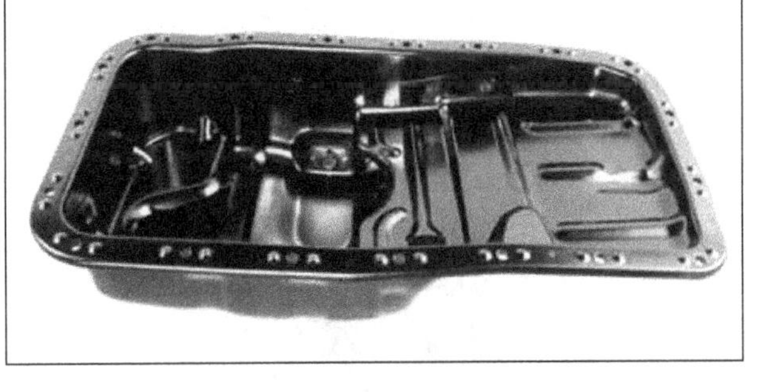

At the bottom of the oil sump, a drain plug is provided to drain out the dirty oil at the time of oil replacement. Generally, the sump is made of pressed steel sheet or aluminium alloy casting is used. The various functions of the oil pan as follows:

- To store the oil for the engine lubrication system.
- Oil pan used to collect the return oil draining.
- To serve as a container for impurities or foreign matters.
- Oil pan provides for cooling of the hot oil in the sump.

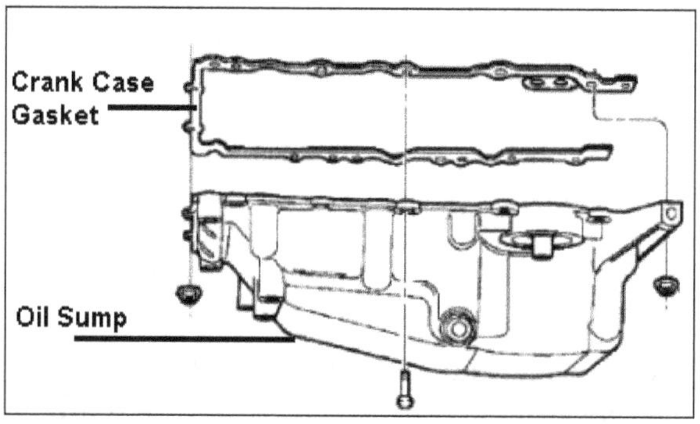

- Working:
    - The oil pump in the lubricating system draws oil from the oil pan and sends it to all working parts in the engine.
    - The oil drains off and runs down into the pan.
    - Thus there is a constant circulation of oil between the pan and the working parts of the engine.

Manifolds

There are separate sets of pipes attached to the cylinder head which carry the air-fuel mixture and the exhaust gases, these are called manifolds. It is generally made of cast iron so that it is able to withstand the high temperature of the exhaust gases.

- Construction:
    - It consists of the air intake, throttle body, intake manifold flange for tail-pipe and flange for a carburettor.
- Working:
    - The air goes into the air intake travels through, throttle body into the intake manifold and from there it goes into the engine through the cylinder head.
    - The inlet manifold carries the air-fuel mixture from the carburettor to the cylinders.
    - The exhaust manifold is the set of pipes carrying exhaust gases from the cylinder head to the exhaust system.

Gaskets

These are used to provide a tight fitting joint between two surfaces.

- Gaskets are found in:
    - The joint between the cylinder head and the cylinder block.
    - Between crankcase and oil pan.
    - Between the cylinder block and manifold.
- Materials used for gaskets are:
    - Cork.
    - Asbestos.
    - Rubber.
- Requirement/properties of the gasket as follows:
    - Conformity: The gaskets should conform to the mating surfaces which may have roughness or warpage.
    - Resistance: It should have resistance to high pressures, extreme temperature and vibrations.

  - Impermeability: The gasket must be impermeable to the fluid.
  - Resistance to Chemical Attack: the gasket should have resistant to the chemicals such as fuel, products of combustion, coolant and engine oil.
  - Provision of Apertures: The gasket must have apertures for any studs, bolts, opening etc.
- Gaskets produced by the fuel-pro USA as follows:
  - Cylinder head gaskets.
  - Oil pan gaskets.
  - Manifold gaskets.
  - Pump gaskets.
- Types of gaskets used in engines:
  - Copper-asbestos gasket.
  - Steel-asbestos gasket.
  - Steel-asbestos-copper gasket.
  - Single steel ridged or corrugated gasket.
  - Stainless steel gasket.

Cylinder Liners

These are cylindrical shapes used in the cylinders to avoid the problem of cylinder wear. It is one of the most important functional parts to make up the interior of an engine.

Cylinder Liners

These can be replaced after they are worn out. These are made of special alloy iron containing silicon, manganese, nickel and chromium. Usually, these are cast centrifugally. These liners resistance to wear and corrosion. These liners are of the oil hardening type and offer considerably longer life for the engine.

- Cylinder liners are of two types: Dry liners and wet liners.
    - Dry liners:

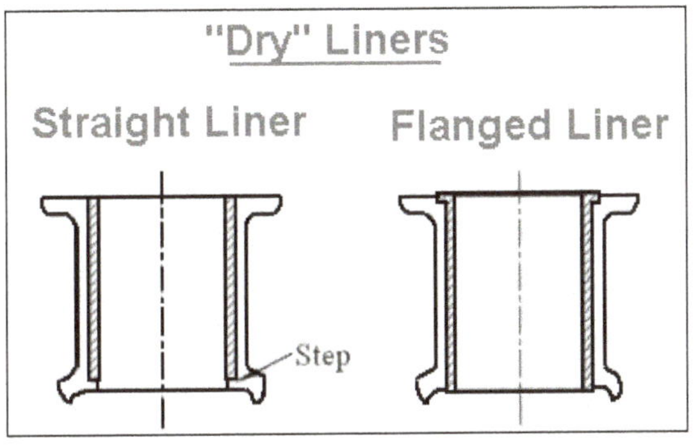

Construction: The construction of a dry liner is shown in the figure. This liner is made in the shape of a barrel with a flange at the top which keeps it into position. The entire outer surface bears against the cylinder block casting and hence these are it be machined accurately at both outer and inner faces. The liner should not be too loose, otherwise, the heat dissipation becomes poor because of the absence of good contact with the cylinder block.

- Wet liners: The figure shows a simple sketch of the wet liner. These liners will be in direct contact with the cooling water at their outer face. Thus, these liners need not be machined very accurately at the outer surface. However, they have been machined accurately at the inner surface. They are resisting corrosion with continuous contact with cooling water and they coated with aluminium at their outer surface.

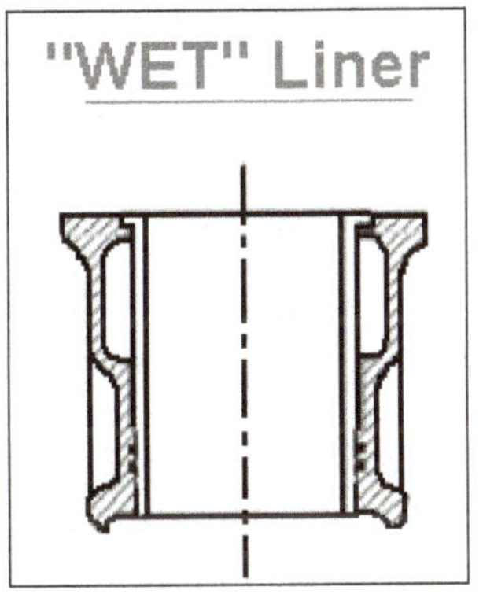

Construction: At the top, the liner is provided with a flange which fits into the groove in the cylinder block. At the bottom of the liner is provided with a groove, generally three in number. The middle groove is left empty for drainage for any water that may leak from the upper ring. And in the top and bottom ones are inserted packing ring, made of synthetic rubber.

Comparison of the dry and wet liner:

- Dry Liners:
    - Dry liners may be provided either in the original design or even afterwards.
    - The construction of the cylinder block very complicated. The cooling effect is not very good.
    - Accurate machining of dry liners for perfect contact with the cylinder casting is essential.
    - In this type, it cannot be finished before fitting. A leak-proof joint is not necessary.
- Wet Liners:
    - Wet liners have to be included in the original design. The construction of the cylinder block is simple.
    - The cooling effect is better because the liner will have direct contact with cooling water.
    - Accurate machining is not essential. In this type, they can be finished before fitting.
    - A leak-proof joint should be made between wet liner and cylinder block.

Pistons

Pitons are most important engine parts compared to others. The piston is a cylindrical plug that moves up and down in the cylinder.

- It helps to convert pressure energy obtained by the combustion of fuel into useful mechanical power and it transfers this power to the crankshaft through the connecting rod.
    - The highest position of the piston reaches in the cylinder is called the Top Dead Centre (TDC).

- The lowest position it reaches is called the Bottom Dead Centre (BDC).

- It is provided with pistons ring about 3 to 5 provide a good seal between the cylinder wall and piston. The efficiency and economy of the engine primarily depend on the working of the piston. The material used for the piston is mainly as follows:
  - Cast iron.
  - Aluminium alloy.
  - Nowadays aluminium alloys are widely used. It may be either cast or forged.

- The piston must possess the following qualities:
  - Rigidly to withstand high pressure.
  - Light in weight, to reduce the reciprocating mass to perform at higher engine speed.
  - Good heat conductivity.
  - Less noise while operating.

- Piston clearance: The piston is usually small in diameter than the bore of the cylinder. The space between the cylinder and the cylinder wall is called the piston clearance. This piston clearance provides a space for a layer of lubricant between the piston and cylinder wall to reduce friction. Generally, piston clearance is 0.025mm to 0.100mm.

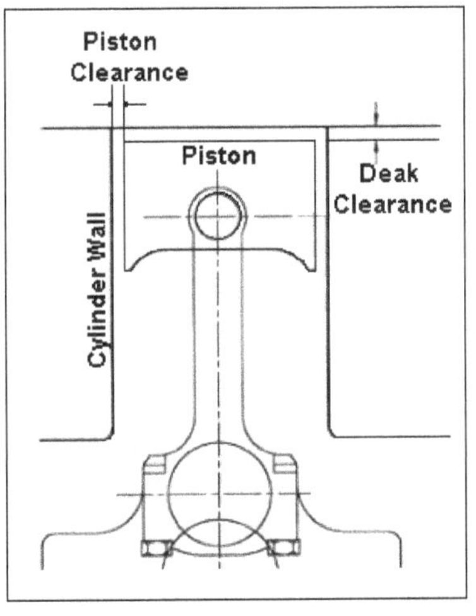

Proper clearance should be maintained between the piston and the cylinder wall.

- If the clearance is too small, there will be a loss of power from excessive friction, more wear, seizing of the piston in the cylinder.

- If the clearance is too much, the piston slap will result. Piston slap means sudden tilting of the cylinder as the piston moves down during the power stroke.

It prevents piston seizure due to high temperature. If there is on clearance then it is not possible to reciprocate piston inside the cylinder.

- Functions of piston: Some of the important function of the piston as follows:
    - To transmit the power developed by fuel combustion to the crankshaft through the connecting rod.
    - To form a seal so that high-pressure combustion gases do not escape to the crankcase.
    - Piston serves as a support for the small end of the connecting rod.
    - To suck the charge and push out the exhaust gases.
- Constructional Features: The cross-section of a piston is shown in the figure.
    - The top of the piston is called the head or crown.
    - Towards the top of the piston, a few grooves are cut to house the piston rings. The bands left between the grooves are known as lands.
    - The part of the piston below the ring is called Skirt is provided with bosses on the inside to support the piston pin (Gudgeon pin).
    - The distance between the axis of the piston pin and the top of the piston crown is called compression height.
- Type of Pistons: The various types of pistons are classified depending on the shape, design, operation. The important types of the piston as follows.
    - Cast Iron Pistons
    - Forged Pistons
    - Cast Steel Piston
    - BI-metal Piston
    - Two-piece Piston
    - Oil-cooled Pistons
    - Anodized Piston
    - Tinned Pistons
- Piston materials: Cast Iron, Aluminium, Lo-Ex Alloy, Invar, Steel alloy. Protective coating: Cadmium plating, Anodised pistons, Tinned pistons, Chromium plating.

Piston Rings

The piston rings are fitted into the grooves of the piston to maintain a good seal between the piston and the cylinder wall. The number of piston rings used is about 2 to 4 compression rings and 1 to 2 oil control ring was used but in modern design the number if rings usually three out of which one is the oil control ring.

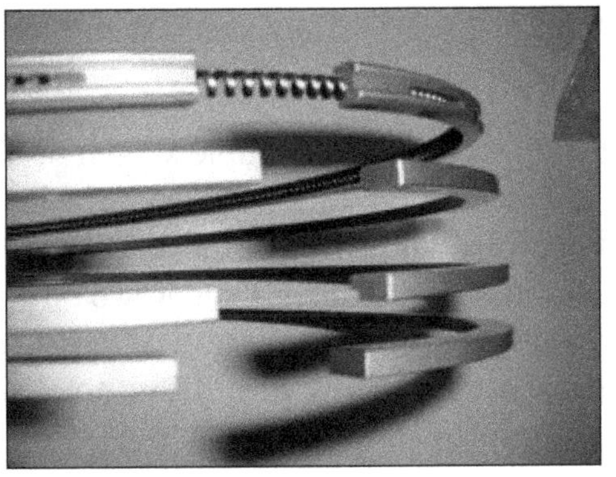

- The function of piston rings are as follows:
    - To form a seal for the high pressures gases from the combustion chamber entering into the crankcase.
    - The piston ring provides easy passage for heat flow from the piston crown to the cylinder walls.
    - To maintain sufficient lubrication oil on cylinder walls throughout the entire length the piston travel, hence it minimizes the cylinder wear.
- Construction: The figure shows the piston ring construction:
    - The ring is generally cast individually and machined carefully so that when in the position it is able to exert uniform pressure against the cylinder walls.
    - A gap has been cut at the end.
    - In practice, the piston ring end gap when installed is kept about 0.30 to 0.35mm.
    - The gap is almost closed when the piston is inside the cylinder, so that piston and cylinder.
- The piston ring end gap may be as follows:
    - Butt type.
    - Taper type.
    - Lap type.
- Material for piston rings: The material generally used for piston rings is fine-grained alloy cast iron containing silicon and manganese. It has good heat and wears resisting qualities. Chromium plated rings are also used for the top ring, which is subjected to the highest working temperatures and the corrosive action of the combustion products.
- Types of Piston Rings: Mainly there are two types of piston rings as follows:
    - Compression rings: Figure shown a simple sketch of Compression rings. These rings effectively seal the compression pressure and the leakage of the combustion gasses.

These are fitted in the top grooves. They also transfer heat from the piston to the cylinder walls.

- Oil control rings: The Figure shows a simple sketch of oil control ring. The main purpose of the oil ring is to scrape the excess oil from the liner and return it back to the oil sump during the downward and upward movement of the piston. It prevents the oil from reaching the combustion chamber. One of two oil control rings is used in a piston. If two rings are used one has fitted above and other is fitted below the gudgeon pin in the piston. These rings are provided with drain holes or slots. These slots allow the scraped oil to reach into the oil sump through the piston holes.

Connecting Rod

Figure showed a connecting rod. It is fitted in between the piston and crankshaft. The main function of the connecting rod is to convert the reciprocating motion of the piston into the rotary motion of the crankshaft. It must be light and strong enough to withstand stress and twisting forces.

- Construction:

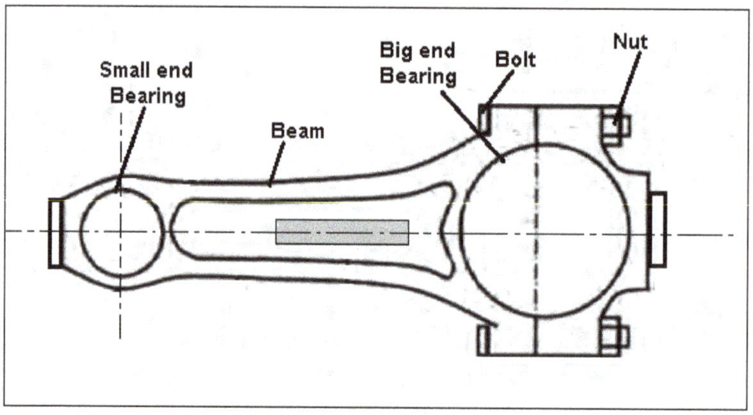

Connecting rod

- The connecting rod usually has I-beam cross-section and is made of alloy steel of duralumin by drop forging.
- Nowadays it is also cast from malleable or spheroidal graphite C.I.

- The small end of the connecting rod has either a solid eye used to connect the piston by the piston pin.
- The big end of the connecting rod is always split is used to connect the crank pin of the crankshaft.

Piston Pin

The piston pin is also called wrist pin or gudgeon pin. It is used for connecting the small end of the connecting rod and the piston.

Piston pin

- Construction: It is made hollow to reduce weight and it is made from case hardened steel.
- Mainly there are three types of piston pins as follows:
  - Set screw types piston pin.
  - Semi-floating piston pin.
  - Fully floating piston pin.

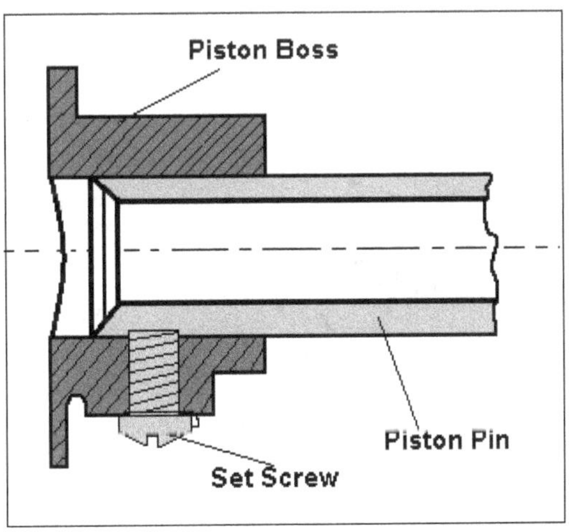

Set screw type piston pin

Figure shows set screw type piston pin, this pin is fastened to the piston to the piston by a set screw such that the connecting rod end swivel has required by the combined reciprocating and rotary motion of the piston and crankshaft.

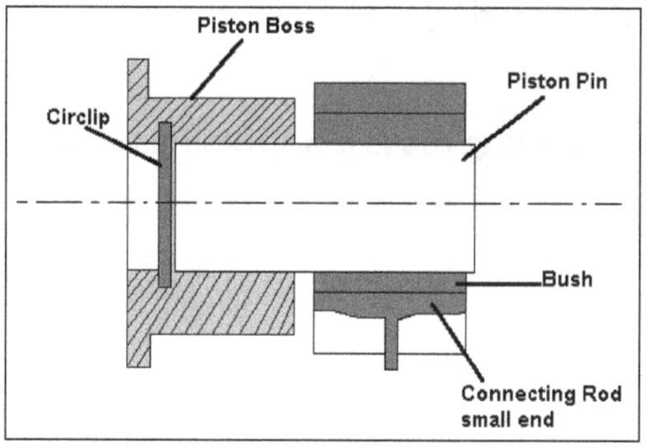

Semifloating piston pin

Figure shows the semi-floating piston pin, it is fastened to the connecting rod with a clamp screw.

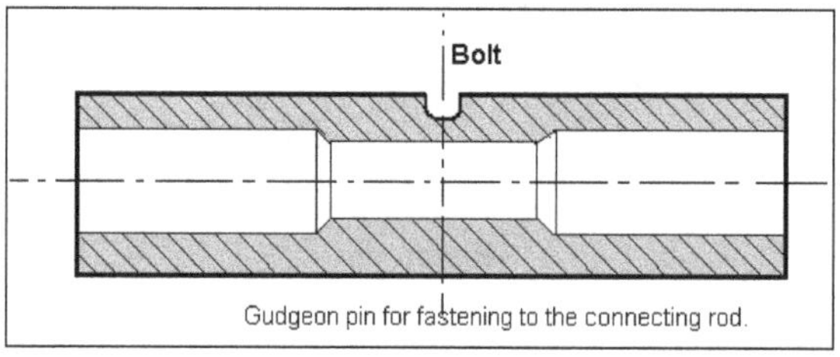

Fully floating pin

Figure shows fully floating piston pin. The pin floats in both the piston bosses and the small end of connecting rod. It is prevented from coming in contact with the cylinder wall by two circlips.

Crankshafts

The crankshaft is the engine component from which the power is taken. It is one of the main power transmission sources in all engine parts. The Crankshaft is the first part of the power transmission system in which the reciprocating motion of the piston is converted into the rotating motion with the help of connecting rod.

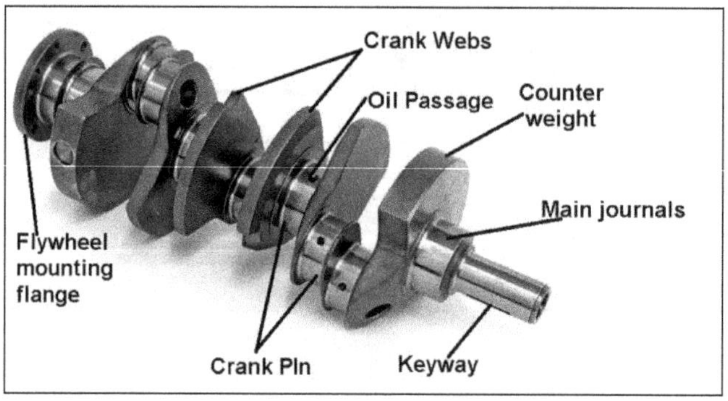

- Construction:
    - The crankshaft is made of casting or forging of heat treated alloy steel and is machined.
    - A crankshaft consists of crankpins, weds, balancing weight and main journals and oil holes.

- The big end of the connecting rod is connected to the crankpin of the crankshaft.
- Centre to centre distance between the crankpin and crankshaft is half of the piston displacement during the stroke.
- Thus one complete revolution of the crankshaft makes two strokes of the piston.

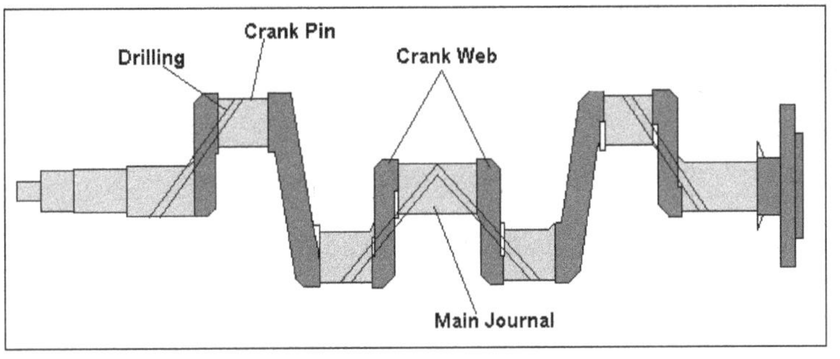

Main bearing journals

- The parts of the crankshaft inside the main bearing are called main journals.
    - Balancing weights are provided on the opposite side wed for balancing. The crankshaft has drilled oil passages through which oil flow the main bearing to the connecting rod bearings.
- The front end of the crankshaft carries three devices that:
    - A gear that drives the camshaft.
    - The vibration damper to control torsional vibration.
    - The fan belt pulley. This pulley drives the engine fan, water pump, and generator with a V-belt.

The rear end of the crankshaft carries flywheel. The flywheel tends to keep the crankshaft running at constant. Next, to the rear end, the main journal and oil seal is fitted. In some engine, oil return threads are provided which return the lubricating oil to the sump.

- The crankshafts are generally of two types:
    - In one piece type, all the parts are integral and are formed by drop forging and then machining.
    - In build-up type, the crankpins and journals are fastened to the crank webs.

Camshaft

A camshaft is a shaft on which cams are mounted. A cam is a device that changes the rotary motion of the camshaft into the linear motion of the follower. A camshaft is responsible for the opening of the valves.

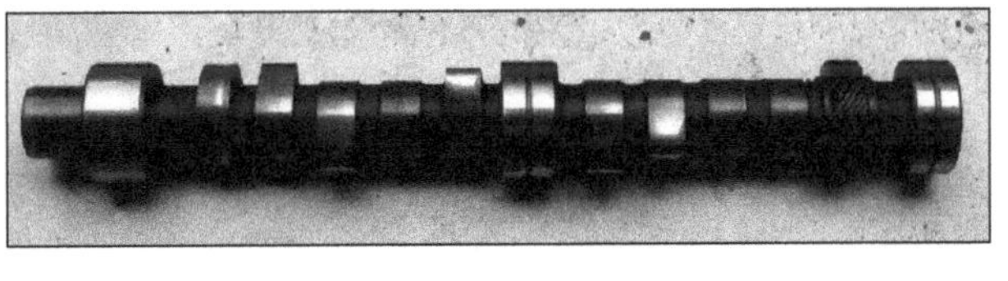

- Construction:
    - A camshaft has a number of cams along the length, two cams for each cylinder, one to operate the inlet valve and the other the exhaust valve.
    - In addition, the camshaft has an eccentric to operate the fuel pump and gear to drive the ignition distributor and oil pump.
    - The camshaft is driven by the crankshaft. The camshaft gear has twice as many teeth as the gear on the crankshaft.
    - Camshaft made from forged alloy steel.

This gives 1:2 gear ratio, the camshaft turns at half the speed of the crankshaft.

- Working:
    - Thus, every two revolutions of the crankshaft produce one revolution o the camshaft and one opening and closing of each valve, in the four-cylinder engine.
    - Thus there is correct opening and closing of the valves takes in relation to the position of the piston in the cylinder.
- There are three types of camshaft drive mechanism as follows:
    - Gear drive.
    - Chain drive.
    - Belt drive.

Flywheel

The flywheel used in a transmission system of a vehicle.

- Construction:
    - A flywheel is a heavy steel wheel attached to the rear end of the crankshaft.
    - The size of the flywheel depends upon the number of cylinders and the construction of the engine.
- Working:
    - During the power stroke, the engine tends to speed up and during the other strokes, it tends to slow down.

- The inertia of the flywheel tends to keep the running of the crankshaft at a constant speed. Hence the engine speed is maintained constant.

Engine Valves

Engine valves are essential to control the timing of air-fuel mixture entry into the cylinder and combustion products out of the cylinders.

- Construction:
    - These are located at the inlet and outlet opening of the engine cylinder.
    - The valves fit on the valve seats in their closed position.

There are three types of engine valves as follows:

- Poppet valve.
- Sleeve valve.
- Rotary valve.

1. Poppet Valve: This is the most widely used valve in automobile engines. The poppet valve is given the name because of its motion of popping up and down. Its construction is very simple. This is also called a mushroom valve, because of its shape.

- Construction:

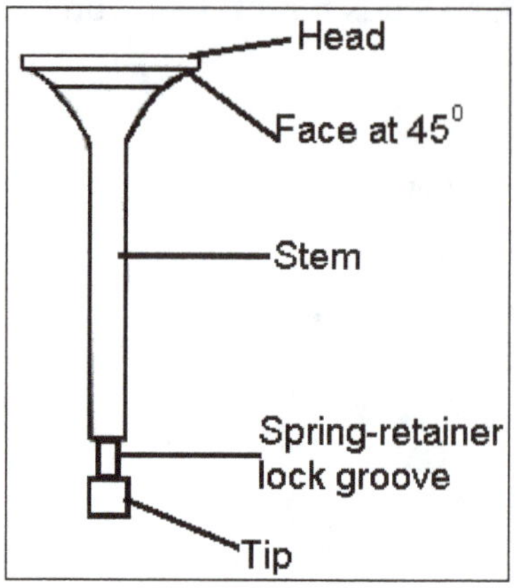

  - It consists of a head and a stem. The valve face usually with an angle of 30° to 45° is ground perfectly, since it has to match with the valve seat for perfect sealing.
  - The stem has a spring retainer lock groove and the stem end is in contact with cam for up and down movements of the valve.

2. Sleeve Valve: The sleeve valve as the name implies, that it is a tube or sleeve kept between the cylinder wall and the piston.

- Construction: The inner surface of the sleeve actually forms the inner cylinder barrel in which the piston slides. The sleeve is in continuous motion and admits and drives out the gases by virtue of the periodic coincidence of port cut in the sleeve with ports formed through the main cylinder casting.
- Advantages:
  - Simple in construction.
  - Sleeve valves are silent in operation.
  - There is noise because there is no noise-making parts like valve cams, racker arm, tappets valves etc.

- The tendency of detonation is less.
- Cooling is very effective as the valve is in contact with water jackets.

- Disadvantages:
    - High oil consumption for lubrication as a larger area of sleeve surface to be lubricated.
    - Cleaning of the ports and the valve is complicated.

3. Rotary Valve: Figure showed a simple sketch of rotary valve. There are many types of rotary valves. The figure shows the disc type rotary valve. It consists of a rotating disc which has a port. While rotating, it communicates s alternately with the inlet and exhaust manifolds.

- Advantages:
    - Rotary valves are simple in construction.
    - These valves are manufactured at cheaper costs.
    - They are suitable for high-speed engines.
    - Stresses and vibrations are less compared to poppet and sleeve valves.
    - They are smooth in operation and are uniform and noise-free operation.
- Disadvantages:
    - It is difficulties in pressure sealing between the rotary disc and cylinder.
    - Efficient valve lubrication is difficult.
- Materials for Valves: The materials used for inlet and exhaust valve are generally different because of the different operating conditions to which valves are subjected. Silico-chrome steel is the material generally used for inlet valves. For exhaust valves molybdenum as added to the silico-chrome. The recent materials for exhaust valve are austenite steel and precipitation hardening steel is generally used.

Governor

In petrol engines, the carburettor control both air and fuel supply to the engine cylinder under speed and load conditions. They vary the supply of air-fuel mixture to meet the given condition. But in the diesel engine, the governor is used to keep the engine speed within limits.

- The main functions of a governor are to regulate the supply of fuel through some mechanism so that the engine speed remains within its range.
- Working:
    - On increased load, the engine speed decreases.
    - When the load decreases, the engine speed increases.

Without a governor, the engine speed increases at lighter loads and the dynamic stresses damage the engine parts. The governor which is set for a particular engine speed operates a mechanism such that more fuel is injected to increase the engine power. Governor, in this case, operates the mechanism to reduce the supply of fuel in the engine. It is essential to keep the engine speed within limits.

- Types of Governor:
  - Mechanical Governor or Torque control Governor or Centrifugal governor.
  - Pneumatic Governor.
  - Hydraulic Governor.

Pushrods

Cylindrical pushrods translate motion from the cam to the valve train in overhead valve (OHV) engines. It is vital that the pushrod faithfully reproduces the desired motion from the cam to the valves. High pushrod stiffness is required to control bending, a low mass is needed to reduce lofting, and good damping is essential to limiting vibration. A fiber reinforced aluminium MMC is used in pushrods for high-performance OHV racing engines. 60% of Nextel 610 Al_2O_3 fibers are used as reinforcement and a base is made of aluminium alloy. The company 3 M Corporation was a first manufacturer of these parts. The aluminium MMC pushrods provide 25% higher bending stiffness and twice the damping capacity relative to 4340 steel pushrods. The MMC density is less than half that of steel, so that engine speed can be increased by an average of 250 to 400 rpm before valve bounce begins. Furthermore, it has been projected that the use of aluminium MMC pushrods could provide an extension in spring life of up to 600% as a result of the improved damping characteristics of the MMC. Service life of these pushrods is far greater than steel ones.

Timing Chains

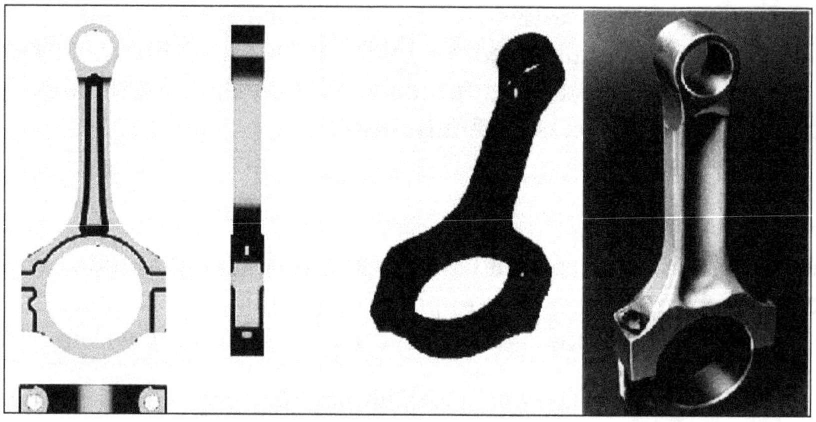

Connecting rods from aluminium MMC.

In the internal combustion engine application, the timing belt connects the crankshaft to the camshaft(s), which in turn controls the opening and closing of the engine's valves. Gear or chain systems can also be used to connect the crankshaft to the camshaft at the correct timing. However gears and shafts constrain the relative location of the crankshaft and camshafts. Chain gear was originally used and 1962, the timing belt was for the first time built in an automobile, Glas 1004.

Engine of Glas 1004 was the first engine where the timing belt was utilized as the camshaft drive (Synchroflex). Since then, both the drives are applied. Initially timing belts were made of rubber, and today, materials as neoprene, urethane, and polyurethane are increasingly applied, and traction element is made from steel, fiberglass, and polyester. Materials for making pulley are steel, cast iron, plastics, and recently aluminium alloys AlCuMgPb, AlZnMgCu0.5-F48, AlCuMgSi (En Aw2017A).

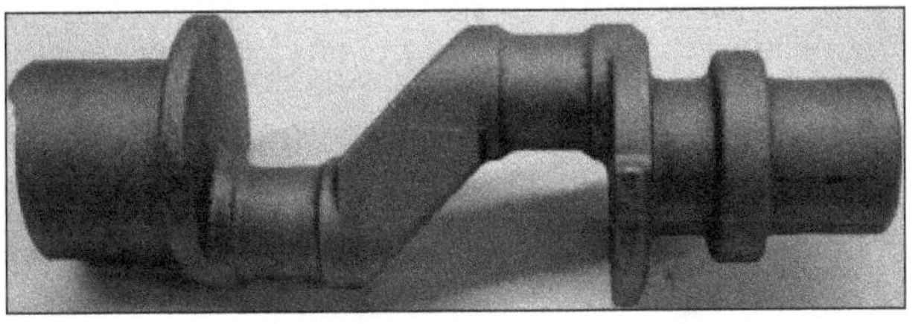

Aluminum/SiC-graphite crankshaft.

Different types of steel are mainly used to make a chain and chainring. Most commonly used are hardened steels (C15, CK15, 17Cr3, 16MnCr5) and tempered steels (C45, C60, 40Mn4, 34Cr4, 41Cr4, 50Mn7).

Timing Cover

Timing Belt Cover

A timing cover is a component typically found in internal combustion engines and is designed to protect components such as the timing gears and chain or belt. In applications that use a timing chain, the timing cover usually seals the front end of the block, while engines with timing belts typically have timing covers that are less integral. Timing covers can be made from either metal or plastic. Engines with timing chains usually have a metal cover, while those with a timing belt may utilize covers made from a variety of different materials.

The timing gear and belt or chain in a car are protected by a timing cover.

Timing chain covers are typically designed to seal the front end of the engine block. This is generally so the timing gears and chain can be lubricated with engine oil. These covers will usually be sealed to the block with a timing cover gasket, which may be responsible for engine oil leaks if it fails. Additionally, the crank shaft will typically pass through a seal set into the timing cover that may potentially leak. Replacing a timing cover gasket is typically a time intensive job, while replacing a crankshaft oil seal is sometimes possible without removing the cover itself.

Engines that utilize a timing chair, rather than a timing belt, typically have metal covers.

Timing covers are also used in conjunction with timing belts. In these applications the front of the engine will typically be sealed with oil seals that are set into the block or head where the crank and cam shafts extend. With the engine otherwise sealed, the function of a timing belt cover is simply to protect the belt from damage. These covers may be constructed of either metal or plastic and may be comprised of several different pieces. Multi-piece timing belt covers may even contain both metal and plastic pieces.

Car and truck engines use many types of sealing gaskets, including to seal the timing cover.

Timing cover gaskets are also often used with timing belt covers, though they are designed to keep contaminants away from the belt. If an oil leak develops at a crank or cam shaft seal, oil may contaminate the timing belt and require its replacement. Oil may also contaminate the cover gasket, causing it to swell and become unusable. In these cases, the seal, belt, and gasket are often replaced at the same time. In some cases, timing belt covers will use special timing cover bolts. These bolts will typically have a thick shoulder that allows them to tighten without cracking the plastic cover. In other cases the cover may have metal bezels in the bolt holes that are designed to perform the same function.

References

- Basic-engine-components-engine-parts: theengineerspost.com, Retrieved 09, January 2020
- Automotive-Engine-Materials-295009898: researchgate.net, Retrieved 29, July 2020
- What-is-a-timing-cover: infobloom.com, Retrieved 06, March 2020
- Carburetor-parts-types-working-function: learnmechanical.com, Retrieved 14, May 2020
- What-is-fuel-injector: mechanicalbooster.com, Retrieved 25, August 2020

2

Internal and External Combustion Engine

There are two main types of heat engines: internal combustion engine and external combustion engines. Internal combustion engines are further divided according to their design, fuel used, cycle of operation, number of strokes, ignition, etc. while external combustion engines are divided into steam and stirling engines. All these diverse types of engines have been carefully analyzed in this chapter.

The classification of the engines depends upon the types of fuel used, cycle of operation, number of stroke, type of ignition, number of cylinders, arrangement of cylinders, valve arrangement, types of cooling etc. these engines are used in different areas such as in automotive industries, aircraft industries, marine industries, etc. according to their suitability they are used in different areas. Basically the engines are of two types, and these are external combustion engines and internal combustion engines.

Internal Combustion Engine

Internal combustion engines (ICE) are the most common form of heat engines, as they are used in vehicles, boats, ships, airplanes, and trains. They are named as such because the fuel is ignited in order to do work inside the engine. The same fuel and air mixture is then emitted as exhaust. This can be done using a piston (called a reciprocating engine), or with a turbine.

Internal Combustion Engine Parts and their Function

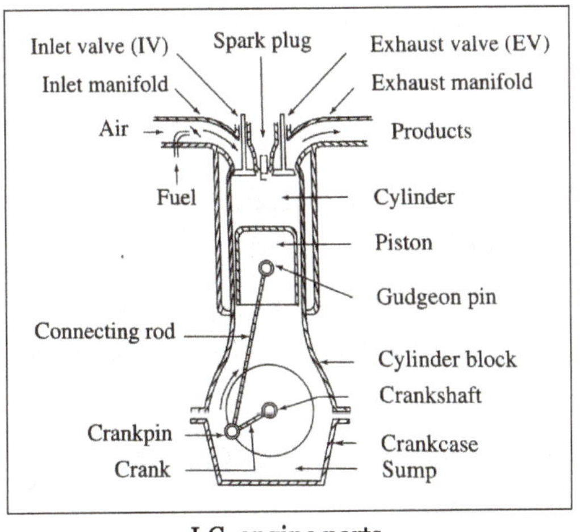

I.C. engine parts

- Cylinder: It is a container fitted with piston, where the fuel is burnt and power is produced.
- Cylinder Head/Cylinder Cover: One end of the cylinder is closed by means of cylinder head. This consists of inlet valve for admitting air fuel mixture and exhaust valve for removing the products of combustion.

- Piston: Piston is used to reciprocate inside the cylinder. It transmits the energy to crank-shaft through connecting rod.
- Piston Rings: These are used to maintain a pressure tight seal between the piston and cylinder walls and also it transfer the heat from the piston head to cylinder wall.
- Connecting Rod: One end of the connecting rod is connected to piston through piston pin while the other is connected to crank through crank pin. It transmits the reciprocatory motion of piston to rotary crank.
- Crank: It is a lever between connecting rod and crank shaft.
- Crank Shaft: The function of crank shaft is to transform reciprocating motion in to a rotary motion.
- Fly wheel: Fly wheel is a rotating mass used as an energy storing device.
- Crank Case: It supports and covers the cylinder and the crank shaft. It is used to store the lubricating oil.

Terminology of ICE

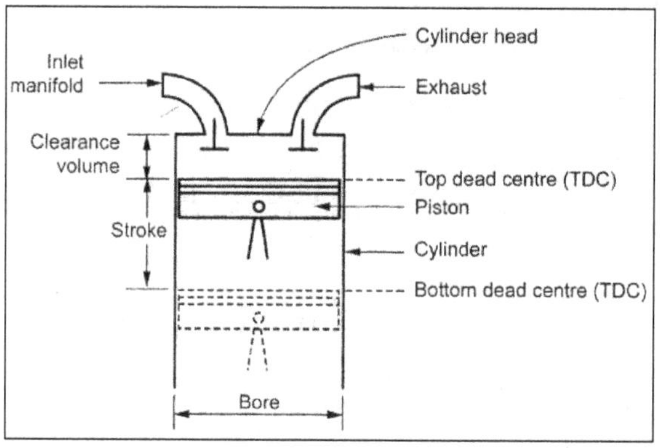

- Bore: The inside diameter of the cylinder is called the bore.
- Stroke: The linear distance along the cylinder axis between the two limiting positions of between the piston is called stroke.
- Top Dead Centre (T.D.C): The top most position of the piston towards cover end side of the cylinder" is called top dead centre. In case of horizontal engine, it is called as inner dead centre.
- Bottom Dead Centre (B.D.C): The lowest position of the piston towards the crank end side of the cylinder is called bottom dead centre. In case of horizontal engine, it is called outer dead centre (O.D.C).
- Clearance Volume: The volume contained in the cylinder above the top of the piston, when the piston is at the top dead centre is called clearance volume.
- Compression Ratio: It is the ratio of total cylinder volume to clearance volume.

According to Design

Reciprocating Engine

Reciprocating internal combustion engines are devices that convert the chemical energy contained in a hydrocarbon into mechanical energy (rotation of a shaft with a certain speed and torque) and into the thermal energy of the waste gases that escape into the atmosphere. Reciprocating internal combustion engines are a well-established and widely used technology. Worldwide production for reciprocating internal combustion engines is over 200 million units per year. Reciprocating engines include both diesel and spark-ignition configurations. They are important for both transportation and for stationary uses. Their sizes range from fractional horsepower engines to 5-story tall marine propulsion systems weighing over 5 million pounds and producing over 80 megawatts (MW) of power. The long history of technical development and high production levels have contributed to making reciprocating engines a rugged, reliable, and economic choice as a prime mover for CHP applications. Reciprocating engine technology has improved dramatically over the past three decades, driven by economic and environmental pressures for power density improvements (more output per unit of engine displacement), increased fuel efficiency, and reduced emissions. Electronic Power Control Modules (PCMs) have made possible more precise control and diagnostic monitoring of the engine process. Stationary engine manufacturers and worldwide engine R&D firms continue to drive advanced engine technology, including accelerating the diffusion of innovative technology and concepts from the automotive market to the stationary market. The features that have made reciprocating engines a leading prime mover for CHP and other distributed generation applications are summarized in table.

Table: Reciprocating engine characteristics.

Size range	Reciprocating engines are available in sizes from 10 kW to over 18 MW.
Thermal output	Reciprocating engines can produce hot water, low pressure steam, and chilled water (using an absorption chiller).
Fast start-up	The fast start-up capability of reciprocating engines allows timely resumption of the system following a maintenance procedure. In peaking or emergency power applications, reciprocating engines can quickly supply electricity on demand.
Black-start capability	In the event of an electric utility outage, reciprocating engines require minimal auxiliary power requirements. Generally only batteries or compressed air are required.
Availability	Reciprocating engines have typically demonstrated availability in excess of 95 percent in stationary power generation applications.
Part-load operation	The high part-load efficiency of reciprocating engines ensures economical operation in electric load following applications.
Reliability and life	Reciprocating engines have proven to be reliable power generators given proper maintenance.
Emissions	Diesel engines have relatively high emissions levels of NO_x and particulates. However, natural gas spark ignition engines have improved emissions profiles.

Applications

Reciprocating engines are well suited to a variety of distributed generation applications, and are used throughout industrial, commercial, and institutional facilities for power generation and CHP. Reciprocating engines start quickly, follow load well, have good part load efficiencies, and generally have high reliabilities. In many cases, having multiple reciprocating engine units further increases overall plant capacity and availability. Reciprocating engines have higher electrical efficiencies than gas turbines of comparable size, and thus lower fuel-related operating costs. In addition, the upfront costs of reciprocating engine gensets are generally lower than gas turbine gensets in sizes below 20 MW. Reciprocating engine maintenance costs are generally higher than comparable gas turbines, but the maintenance can often be handled by in-house staff or provided by local service organizations.

Combined Heat and Power

There are over 2,000 active reciprocating engine combined heat and power (CHP) installations in the U.S. providing nearly 2.3 gigawatts (GW) of power capacity. These systems are predominantly spark ignition engines fueled by natural gas and other gaseous fuels (biogas, landfill gas). Natural gas is lower in cost than petroleum based fuels and emissions control is generally more effective using gaseous fuels. Reciprocating engine CHP systems are commonly used in universities, hospitals, water treatment facilities, industrial facilities, and commercial and residential buildings. Facility capacities range from 30 kW to 30 MW, with many larger facilities comprised of multiple units. Spark ignited engines fueled by natural gas or other gaseous fuels represent 84 percent of the installed reciprocating engine CHP capacity.

Thermal loads most amenable to engine-driven CHP systems in commercial/institutional buildings are space heating and hot water requirements. The simplest thermal load to supply is hot water. The primary applications for CHP in the commercial/institutional and residential sectors are those building types with relatively high and coincident electric and hot water demand such as colleges and universities, hospitals and nursing homes, multifamily residential buildings, and lodging. If space heating needs are incorporated, office buildings, and certain warehousing and mercantile/service applications can be economical applications for CHP. Technology development efforts targeted at heat activated cooling/refrigeration and thermally regenerated desiccants expand the application of engine-driven CHP by increasing the thermal energy loads in certain building types. Use of CHP thermal output for absorption cooling and/or desiccant dehumidification could increase the size and improve the economics of CHP systems in already strong CHP markets such as schools, multifamily residential buildings, lodging, nursing homes and hospitals. Use of these advanced technologies in other sectors such as restaurants, supermarkets and refrigerated warehouses provides a base thermal load that opens these sectors to CHP application. Reciprocating engine CHP systems usually meet customer thermal and electric needs as in the two hypothetical examples below:

- A typical commercial application for reciprocating engine CHP is a hospital or health care facility with a 1 MW CHP system comprised of multiple 200 to 300 kW natural gas engine gensets. Thesystem is designed to satisfy the baseload electric needs of the facility. Approximately 1.6 MW of thermal energy (MW_{th}), in the form of hot water, is recovered from engine exhaust and engine cooling systems to provide space heating and domestic hot water to the facility as well as to drive absorption chillers for space conditioning during summer months. Overall efficiency of this type of CHP system can exceed 70 percent.
- A typical industrial application for engine CHP would be a food processing plant with a 2 MW natural gas engine-driven CHP system comprised of multiple 500 to 800 kW engine

gensets. The system provides baseload power to the facility and approximately 2.2 MW_{th} low pressure steam for process heating and washdown. Overall efficiency for a CHP system of this type approaches 75 percent.

Emergency/Standby Generators

Reciprocating engine emergency/standby generators are used in a wide variety of settings from residential homes to hospitals, scientific laboratories, data centers, telecommunication equipment, and modern naval ships. Residential systems include portable gasoline fueled spark-ignition engines or permanent installations fueled by natural gas or propane. Commercial and industrial systems more typically use diesel engines. The advantages of diesel engines in standby applications include low upfront cost, ability to store on-site fuel if required for emergency applications, and rapid start-up and ramping to full load. Because of their relatively high emissions of air pollutants, such diesel systems are generally limited in the number of hours they can operate. These systems may also be restricted by permit from providing any other services such as peak-shaving.

Peak Shaving

Engine generators can supply power during utility peak load periods thereby providing benefits to both the end user and the local utility company. The facility can save on peak power charges and the utility can optimize operations and minimize investments in generation, transmission, and distribution that are used only 0-200 hours/year.

Rotating Engine

Rotatory engine is an internal-combustion engine in which the combustion chambers and cylinders rotate with the driven shaft around a fixed control shaft to which pistons are affixed; the gas pressures of combustion are used to rotate the shaft. Some of these engines have pistons that slide in toroidal (doughnut-shaped) cylinders; others have single- and multiple-lobed rotors. Early rotary engines were used in World War I aircraft. They were air-cooled, with cylinders arranged circularly around a crankshaft fastened rigidly to the fuselage. The propeller was attached directly to the circular frame on which the rotating cylinders were mounted. Various inefficiencies in these engines led to their abandonment after the war.

After World War II the development of a new kind of rotary engine reawakened interest. The Wankel is the most fully developed and widely used of the rotary engines. In the Wankel engine the rotor, in the form of an equilateral triangle, rotates with an orbital motion in a specially shaped casing and forms rotating crescent-shaped combustion chambers between its sides and the curved wall of the casing. The three apexes of the rotor are provided with spring-loaded sealing plates that maintain a continuous sliding contact with the concave inner surface of the casing, and the combustion chambers increase and decrease successively in size as the rotor turns. The fuel charge from a carburetor enters the chamber through an intake port, is compressed as the size of the chamber is reduced by the rotation of the rotor, and at the appropriate time is ignited by a spark plug. The Wankel engine was first tested for use in automobiles in 1956. It has since come to be used for such industrial applications as driving air compressors, where small, light-weight, high-speed engines with mechanical simplicity are needed.

According to Fuel Used

Petrol Engine

A petrol engine or gasoline engine is a type of heat engine, specifically an internal combustion that is powered by gasoline. These engines are the most common ways of making motor vehicles move. While turbines can be powered by gasoline, a gasoline engine refers specifically to piston-driven gasoline engines. Gasoline engines are a lot of the reason why the world takes so much oil out of the ground to refine into petroleum products like gasoline. Worldwide, transportation is roughly 18% of our primary energy use and gasoline is a little less than half of that. This means that gasoline engines use roughly 8% of the total primary energy of the world.

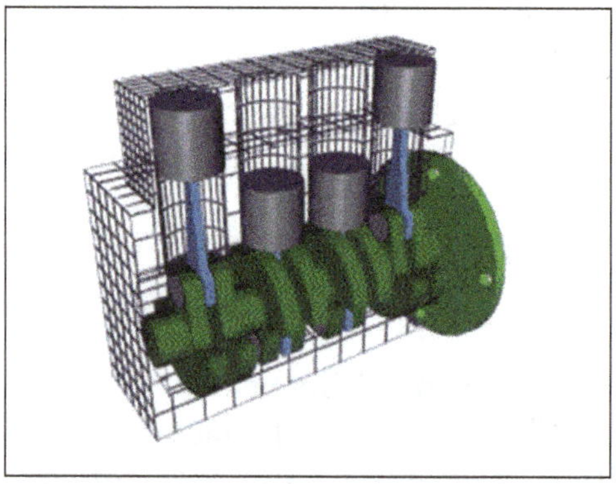

A moving diagram of an inline four engine. The pistons are in gray, the crankshaft is in green, and the block is transparent.

Anatomy of an Engine

- Block: The block is the backbone of the engine. It is a large block of metal, usually aluminium or steel (Formula 1 uses a magnesium alloy), with holes cut into it for the cylinders.
- Cylinders: The cylinders of an engine are where the work is done. Fuel is injected into the cylinders, where it is ignited by the spark plugs, which moves the pistons, doing work.
- Pistons: Pistons are devices that slide up and down inside the cylinders. Their job is to slide in and out, connected to the crankshaft, in order to turn the combusting gasoline into work.
- Spark Plugs: The job of a spark plug is to ignite the fuel inside the cylinder. The rapid expansion of the fuel, due to the heat created, does work on the piston, moving it away from the spark plug.
- Camshaft: A camshaft is a device that manages the timing of the engine. A camshaft's job is to regulate when fuel is let into the engine, and when exhaust is let out.
- Injectors: The purpose of a fuel injector is to atomize the fuel. This means to turn the liquid fuel into a mist, which drastically increases its surface area. This allows the fuel to combust more quickly, giving greater impulse to the piston.
- Crankshaft: The crankshaft is the glue that connects the parts of the engine. Its purpose is to turn the linear (up and down) motion of the pistons into rotational motion. One end of

the crankshaft is attached to the camshaft, via a timing belt. The other end is connected to the flywheel, which regulates the power coming out of the engine, sort of like a surge protector for your computer.

- Flywheel: The flywheel is a power management device for the engine. It is connected to the clutch, which is connected to the transmission.

Diesel Engine

A diesel engine is a type of internal combustion engine that uses compression ignition to turn energy in diesel fuel to rotary mechanical energy. The diesel engine was named after its inventor, Rudolf Diesel, who was born in Paris in 1958. The first diesel engine was prototyped in 1893 and originally designed to run on peanut oil. Diesel fuel was so named because it was used to run diesel engines, diesel engines were not named after the fuel, which at the time of the first diesel engine, was an useless byproduct of extracting paraffin and kerosene from crude oil. In 1894 this waste product was finally named "diesel".

Working of Diesel Engine

A diesel engine works by using pistons to compress a mixture of air (containing oxygen) with diesel fuel. When this air is compressed at a ratio of about 15:1 the mixture explodes forcing the piston back up and creating the reciprocating motion. This motion is then converted to rotary motion by the engines crank shaft.

Major Components in a Diesel Engine

- The Engines Fuel System: The fuel system includes the fuel injection pump, the lift pump, the injectors and all the fuel pipes. There will also be some fuel filters and maybe a water separator preventing poor quality fuel from damaging your diesel engine.
- The Engines Lubrication System/Oil System: The lubrication system keeps your engine running smoothly, preventing the moving parts from wearing by using the oil under pressure to lubricate and reduce friction. The oil system will have an oil pump and oil filters to keep the oil clean from contaminants.
- The Engines Cooling System: The cooling system handles the engine coolant-normally a mixture of distilled water and glycol with some additional additives to help prevent corrosion. There may also be a coolant filter on some engines and a "water pump" which is actually a coolant pump. The coolant pump is used to push the coolant round the engine and whichever device is used to cool the liquid-usually a radiator, but sometimes a heat exchanger.
- The Engines Exhaust System: Getting rid of the waste combustion gas is very important-moving the waste gas from the engines cylinders out through the exhaust manifold to the main muffler system that reduces the noise. The muffler isn't usually part of the engine, but an addition to reduce the noise to the customers' requirements. The exhaust gas will pass through the turbo charger to make it spin where one is fitted.

- The Engines Turbo Charger: Most engines will have a turbo fitted. This device compresses the combustion air to make the engine more powerful.

A diesel engine is made from hundreds of component parts, but what are the major parts of a diesel engine?

- The Engine Block
- The Pistons
- The Crank Shaft
- The Injection Pump and Engine Governing System
- The injectors
- The Starter Motor
- The Head
- The Valves
- Often a Turbocharger
- The Fuel Filters
- The Oil Filters
- The Air Filters
- The Flywheel

Gas Engine

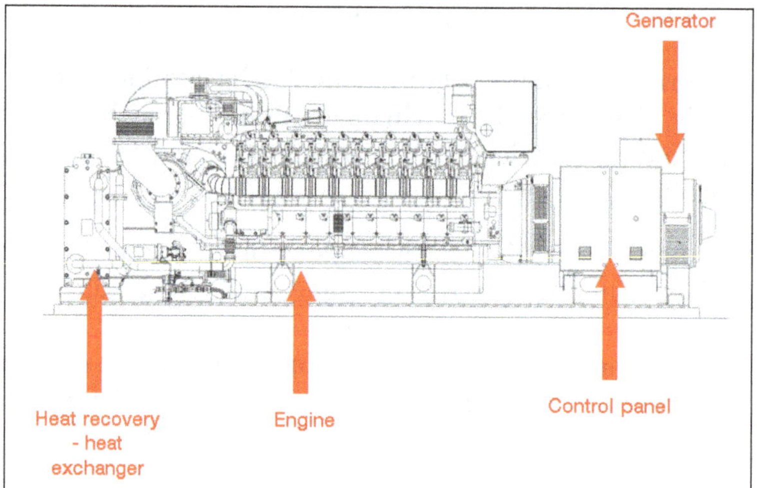

Gas engine basic components

A gas engine is an internal combustion engine which uses blast furnace gas, producer gas, natural gas and others as fuel. The first practical gas engine was built in 1860 by a Frenchman named Lenoir, but as it was largely improved by Dr. Otto, his name was given to its cycle of operations. He made his first gas engine in 1876. The Otto cycle utilizes an ignition source such as a spark or small amount of pilot fuel to start the gas fuel burn. The image below shows the basics of a stationary gas engine and generator used for the production of power. It consists of four main components – the engine which is fuelled by different gases. Once the gas is burnt in the cylinders of the engine, the force turns a crank shaft within the engine. The crank shaft turns an alternator which results in the generation of electricity. Heat from the combustion process is released from the cylinders. This must be either recovered and used in a combined heat and power configuration or dissipated via dump radiators located close to the engine. Finally, and importantly there are advanced control systems to facilitate robust performance of the generator.

Types of Gas Engines

Gas engines can be classified according to the fuel they are most suited to use:

- Biogas Engines.
- Natural Gas Engines (including landfill gas engines).
- Producer Gas Engines.

Of the above, gas engines from renewable energy sources are biogas engines and producer gas engines.

Capacity Range of Gas Engines

Gas engines start from very small capacities (10 kW) to capacities as large as 4 MW. However, the usual capacities of gas engines used are in the range of 50 kW to about 1.5 MW.

Applications for Gas Engines

Typical applications are base load power, including combined heat and power. Gas engines are rarely used for standby applications, which remain largely the province of diesel engines. The natural gas engines (LNG) are getting more into the marine market, as the lean-burn gas engine can meet the new emission requirements without any extra fuel treatment or exhaust cleaning systems. Biogas engines are typically for power production in a variety of industries that have abundant access to organic waste – animal farms, large factories that generate significant amounts of human and animal waste, vegetable markets etc.

Efficiencies of Gas Engines

Gas engines typically have a mechanical efficiency between 35-45%. The best engines can achieve a mechanical efficiency of slightly more than 48%. Large engines are more efficient than small engines. Gas engines running on biogas typically have a slightly lower efficiency (~1-2%) and syngas reduces the efficiency further still. GE Jenbacher's recent J624 engine is the world's first 24 cylinder gas engine with a high efficiency running on methane.

Why are Gas Engines Preferred over Combustion based Steam Turbines for Biomass Power Production?

Biomass power plants typically follow one of the two processes: Combustion or Gasification. The general rule of thumb has been that for power plants smaller than 2 MW, gasification route is considered most appropriate. For those power plants beyond 2 MW, combustion using a steam rankine cycle is considered appropriate. The reasons are not difficult to fathom. While gas engines typically are not available beyond 2 MW, their efficiencies can be as high as 40% for this capacity range. Steam rankine cycles perform poorly at small scales, and offer reasonable efficiencies only beyond 5 MW (in fact their efficiencies increase significantly only beyond 10 MW). As a result, the preferred route (combustion or gasification) for biomass power plants is dependent usually on the capacity of the biomass power plant. Where gasification is the preferred route, gas engines naturally come into the picture. These gas engines are typically those that can run on producer gas, which is the resulting fuel from biomass gasification.

Renewable Energy Sectors that can Use Gas Engines

- Biogas: Industrial sectors producing the following can use biogas engines for power production:
    - Livestock Manures
    - Waste Feed
    - Food-Processing Wastes
    - Slaughterhouse Wastes
    - Milkhouse Wash Water
    - Fresh Produce Wastes
    - Industrial Wastes
    - Food Cafeteria Wastes
    - Sewage Sludge

In addition to the above, other agricultural operations afford opportunities for biogas production. For example, cassava-processing plants, which produce starch, are common in India and may utilize biogas for electric power.

- Producer Gas: The following segments are suited to use producer gas engines for power production:
    - Agricultural Farms – Jute sticks, cotton stalks, coconut shells.
    - Mills -Rice Mills, Oil Mills (groundnut shells).
    - Cotton Mills (cotton stalk).
    - Companies with access to large amounts of waste biomass such as Julieflora and other Woody Biomass.

According to Cycle of Operation

Otto Cycle Engine

The Otto Cycle is named after Nikolaus Otto (1832–1891, German) who is credited as the first creator of a petroleum fuel based internal combustion engine operating under a four-stroke cycle. Otto's engine work during the eve of the industrial revolution was a critical development leading to the dawn of the automotive industry in the late 19th century. The Otto Cycle involves four engine strokes. A stroke is defined as a process in which the cylinder's piston moves up to the top of the engine or down to the bottom of the engine. The cycle begins with the intake stroke in which fresh reactants (fuel and air) are drawn into the engine's cylinder by a downward-expanding piston motion. The second stroke is compression in which the piston moves from the bottom to top of the engine thereby reducing the volume of the fuel and air. Compressing the reactants before burning them allows for greater work and power to be extracted from the following combustion process. When the piston is near the top of its stroke combustion begins and expansion of the burned gases continues into the expansion stroke. The last of the four strokes is the exhaust stroke in which the burned products of combustion are pushed out of the engine in order to make space for the next cycle's intake stroke. The different processes are shown in figure:

- Intake stroke, gasoline vapor and air drawn into engine ($5 \rightarrow 1$).
- Compression stroke, p, T increase ($1 \rightarrow 2$).
- Combustion (spark), short time, essentially constant volume ($2 \rightarrow 3$). Model: heat absorbed from a series of reservoirs at temperatures T_2 to T_3.
- Power stroke: expansion ($3 \rightarrow 4$).
- Valve exhaust: valve opens, gas escapes.
- ($4 \rightarrow 1$) Model: rejection of heat to series of reservoirs at temperatures T_4 to T_1.
- Exhaust stroke, piston pushes remaining combustion products out of chamber ($1 \rightarrow 5$).

We model the processes as all acting on a fixed mass of air contained in a piston-cylinder arrangement, as shown in figure.

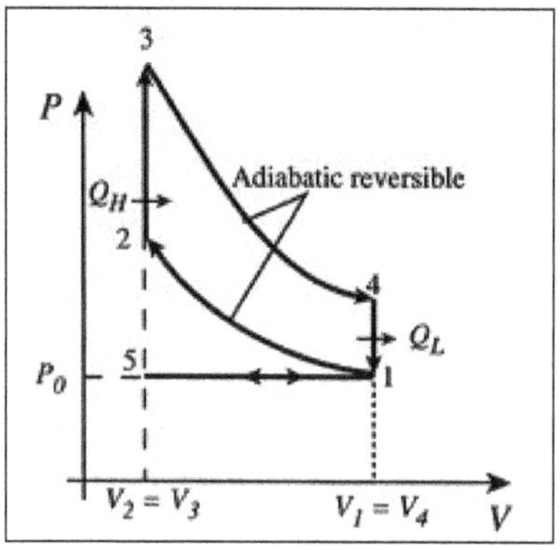

The ideal Otto cycle.

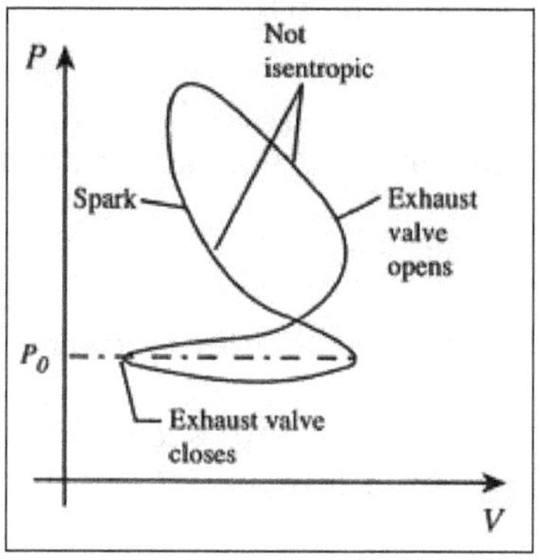

Sketch of an actual Otto cycle.

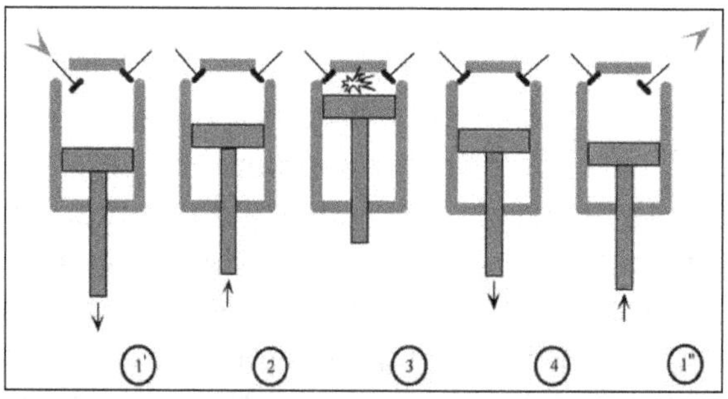

Piston and valves in a four-stroke internal combustion engine.

The actual cycle does not have the sharp transitions between the different processes that the ideal cycle has, and might be as sketched in figure.

Efficiency of an Ideal Otto Cycle

The starting point is the general expression for the thermal efficiency of a cycle:

$$\eta = \frac{\text{work}}{\text{heatinput}} = \frac{Q_H + Q_L}{Q_H} = 1 + \frac{Q_L}{Q_H}$$

The convention, as previously, is that heat exchange is positive if heat is flowing into the system or engine, so Q_L is negative. The heat absorbed occurs during combustion when the spark occurs, roughly at constant volume. The heat absorbed can be related to the temperature change from state 2 to state 3 as:

$$\begin{aligned} Q_H &= Q_{23} = \Delta U_{23} \qquad (W_{23} = 0) \\ &= \int_{T_2}^{T_3} C_v dT = C_v (T_3 - T_2) \end{aligned}$$

The heat rejected is given by (for a perfect gas with constant specific heats):

$$Q_L = Q_{41} = \Delta U_{41} = C_v (T_1 - T_4)$$

Substituting the expressions for the heat absorbed and rejected in the expression for thermal efficiency yields.

$$\eta = 1 - \frac{T_4 - T_1}{T_3 - T_2}$$

We can simplify the above expression using the fact that the processes from 1 to 2 and from 3 to 4 are isentropic:

$$T_4 V_1^{\gamma-1} = T_3 V_2^{\gamma-1}, \qquad T_1 V_1^{\gamma-1} = T_2 V_2^{\gamma-1}$$

$$(T_4 - T_1) V_1^{\gamma-1} = (T_3 - T_2) V_2^{\gamma-1}$$

$$\frac{T_4 - T_1}{T_3 - T_2} = \left(\frac{V_2}{V_1}\right)^{\gamma-1}$$

The quantity $V_1 / V_2 = r$ is called the compression ratio. In terms of compression ratio, the efficiency of an ideal Otto cycle is:

$$\eta_{Otto} = 1 - \frac{1}{(V_1 / V_2)^{\gamma-1}} = 1 - \frac{1}{r^{\gamma-1}}$$

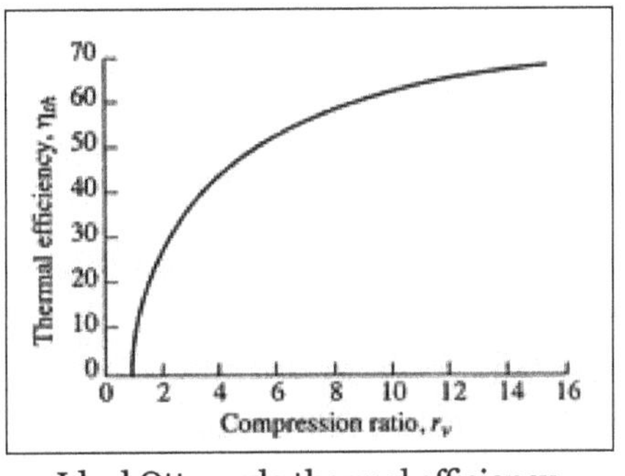

Ideal Otto cycle thermal efficiency.

The ideal Otto cycle efficiency is shown as a function of the compression ratio in figure. As the compression ratio, r, increases, η_{Otto} increases, but so does T_2. If T_3 is too high, the mixture will ignite without a spark (at the wrong location in the cycle).

Engine Work and Rate of Work per Unit Enthalpy Flux

The non-dimensional ratio of work done (the power) to the enthalpy flux through the engine is given by:

$$\frac{\text{Power}}{\text{Enthalpyflux}} = \frac{\dot{W}}{\dot{m} c_p T_1} = \frac{\dot{Q}_{23} \eta_{Otto}}{\dot{m} c_p T_1}$$

There is often a desire to increase this quantity, because it means a smaller engine for the same power. The heat input is given by:

$$\dot{Q}_{23} = \dot{m}_{fuel} \Delta h_{fuel},$$

Where,

- Δh_{fuel} is the heat of reaction, i.e. the chemical energy liberated per unit mass of fuel.
- $\dot{m}_{fuel}$ is the fuel mass flow rate.

The non-dimensional power is

$$\frac{\dot{W}}{\dot{m}c_pT_1} = \frac{\dot{m}_{fuel}}{\dot{m}}\frac{\Delta h_{fuel}}{c_pT_1}\left[1-\frac{1}{r^{\gamma-1}}\right]$$

The quantities in this equation, evaluated at stoichiometric conditions are:

$$\frac{\dot{m}_{fuel}}{\dot{m}} \approx \frac{1}{15}$$

$$\frac{\Delta h_{fuel}}{c_pT_1} \approx \frac{4\times10^7}{10^3\times288},$$

So,

$$\frac{\dot{W}}{\dot{m}c_pT_1} \approx 9\left[1-\frac{1}{r^{\gamma-1}}\right]$$

Diesel Cycle Engine

In the 1890s, a German inventor, Rudolf Diesel has patented his invention of an efficient, slow burning, compression ignition, internal combustion engine. The original cycle proposed by Rudolf Diesel was a constant temperature cycle. In later years Diesel realized his original cycle would not work and he adopted the constant pressure cycle, which is known as the Diesel cycle. Diesel cycle is one of most common thermodynamic cycles that can be found in automobile engines and describes the functioning of a typical compression ignition piston engine. The Diesel engine is similar in operation to the gasoline engine. The most important difference is that:

- There is no fuel in the cylinder at the beginning of the compression stroke, therefore an autoignition does not occur in Diesel engines.
- Diesel engine uses compression ignition instead of spark ignition.
- Because of the high temperature developed during the adiabatic compression, the fuel ignites spontaneously as it is injected. Therefore no spark plugs are needed.
- Before the beginning of the power stroke, the injectors start to inject fuel directly into the combustion chamber and therefore first part of power stroke occurs approximately at the constant pressure.
- Higher compression ratios can be achieved in Diesel engines, than in Otto engines.

The Diesel engine is similar in operation to the gasoline engine. In this picture, there is an Otto engine, which is ignited by a spark plug instead of compression itself.

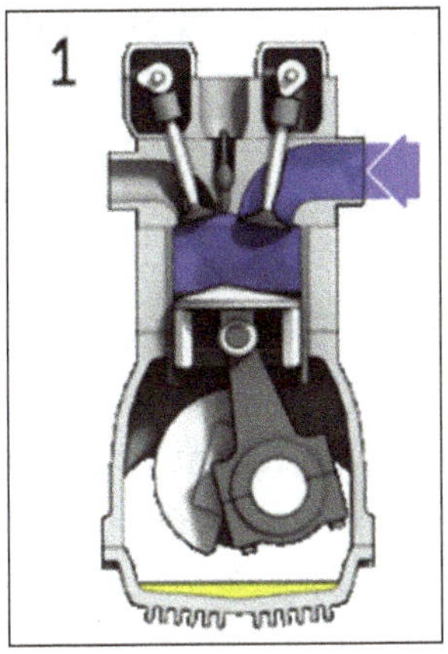

Four stroke engine – Otto engine.

In contrast to Otto cycle, the Diesel cycle does not execute isochoric heat addition. In an ideal Diesel cycle, the system executing the cycle undergoes a series of four processes: two isentropic (reversible adiabatic) processes alternated with one isochoric process and one isobaric process. Since Carnot's principle states that no engine can be more efficient than a reversible engine (a Carnot heat engine) operating between the same high temperature and low temperature reservoirs, the Diesel engine must have lower efficiency than the Carnot efficiency. A typical diesel automotive engine operates at around 30% to 35% of thermal efficiency. About 65-70% is rejected as waste heat without being converted into useful work, i.e. work delivered to wheels. In general, engines using the Diesel cycle are usually more efficient, than engines using the Otto cycle. The diesel engine has the highest thermal efficiency of any practical combustion engine. Low-speed diesel engines (as used in ships) can have a thermal efficiency that exceeds 50%. The largest diesel engine in the world peaks at 51.7%.

Diesel Cycle – Processes

In an ideal Diesel cycle, the system executing the cycle undergoes a series of four processes: two isentropic (reversible adiabatic) processes alternated with one isochoric process and one isobaric process.

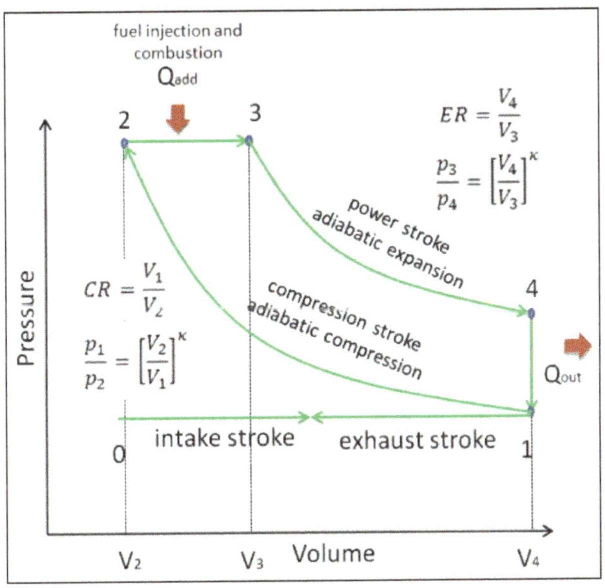

pV diagram of an ideal Diesel cycle.

- Isentropic compression (compression stroke): The air is compressed adiabatically from state 1 to state 2, as the piston moves from bottom dead center to top dead center. The surroundings do work on the gas, increasing its internal energy (temperature) and compressing it. On the other hand the entropy remains unchanged. The changes in volumes and its ratio (V_1/V_2) is known as the compression ratio.

- Isobaric expansion (ignition phase): In this phase (between state 2 and state 3) there is a constant-pressure (idealized model) heat transfer to the air from an external source (combustion of injected fuel) while the piston is moving toward the V_3. During the constant pressure process, energy enters the system as heat Q_{add}, and a part of work is done by moving piston.

- Isentropic expansion (power stroke): The gas expands adiabatically from state 3 to state 4, as the piston moves from V_3 to bottom dead center. The gas does work on the surroundings (piston) and loses an amount of internal energy equal to the work that leaves the system. Again the entropy remains unchanged. The volume ratio (V_4/V_3) is known as the isentropic expansion ratio.

- Isochoric decompression (exhaust stroke): In this phase the cycle completes by a constant-volume process in which heat is rejected from the air while the piston is at bottom dead center. The working gas pressure drops instantaneously from point 4 to point 1. The exhaust valve opens at point 4. The exhaust stroke is directly after this decompression. As the piston moves from bottom dead center (point 1) to top dead center (point 0) with the exhaust valve opened, the gaseous mixture is vented to the atmosphere and the process starts anew.

During the diesel cycle, work is done on the gas by the piston between states 1 and 2 (isentropic compression). Work is done by the gas on the piston between stages 2 and 3 (isobaric heat addition) and between stages 2 and 3 (isentropic expansion). The difference between the work done by the gas and the work done on the gas is the net work produced by the cycle and it corresponds to the area enclosed by the cycle curve. The work produced by the cycle times the rate of the cycle (cycles per second) is equal to the power produced by the diesel engine.

Isentropic Process

An isentropic process is a thermodynamic process, in which the entropy of the fluid or gas remains constant. It means the isentropic process is a special case of an adiabatic process in which there is no transfer of heat or matter. It is a reversible adiabatic process. The assumption of no heat transfer is very important, since we can use the adiabatic approximation only in very rapid processes.

Isentropic Process and the First Law

For a closed system, we can write the first law of thermodynamics in terms of enthalpy:

$$dH = dQ + Vdp$$

or

$$dH = TdS + Vdp$$

Isentropic Process (dQ = 0)

$$dH = Vdp \rightarrow W = H_2 - H_1 \rightarrow H_2 - H_1 = C_p (T_2 - T_1) \text{ (for ideal gas)}$$

Isentropic Process of the Ideal Gas

The isentropic process (a special case of adiabatic process) can be expressed with the ideal gas law as:

$$pV^{\kappa} = \text{constant}$$

or

$$p_1V_1^{\kappa} = p_2V_2^{\kappa}$$

in which $\kappa = c_p/c_v$ is the ratio of the specific heats (or heat capacities) for the gas. One for constant pressure (c_p) and one for constant volume (c_v). Note that, this ratio $\kappa = c_p/c_v$ is a factor in determining the speed of sound in a gas and other adiabatic processes.

Table: Isentropic process – main characteristics.

First Law	$dQ = TdS = 0$ $dU = pdV$ $dH = Vdp$
Ideal Gas Relation	pV^k constant
P, V, T Relations	$\frac{p_1}{p_2} = \left[\frac{V_2}{V_1}\right]^{ê} = \left[\frac{T_1}{T_2}\right]^{\frac{ê}{ê-1}}$
Change in Internal Energy	$dU = mc_V(T_2 - T_1)$
Change in Enthalpy	$dU = mc_p(T_2 - T_1)$
Heat Transfer	0
pdV Work	$W_{dV} = pdV = dU = mc_V(T_2 - T_1)$
Vdp Work	$W_{dp} = Vdp = dU = mc_p(T_2 - T_1)$

Isochoric Process

An isochoric process is a thermodynamic process, in which the volume of the closed system remains constant (V = const). It describes the behavior of gas inside the container, that cannot be deformed. Since the volume remains constant, the heat transfer into or out of the system does not the pΔV work, but only changes the internal energy (the temperature) of the system.

Isochoric Process and the First Law

The classical form of the first law of thermodynamics is the following equation:

$$dU = dQ - dW$$

In this equation dW is equal to dW = pdV and is known as the boundary work. Then:

$$dU = dQ - pdV$$

In isochoric process and the ideal gas, all of heat added to the system will be used to increase the internal energy.

Isochoric process (pdV = 0)

$$dU = dQ \quad \text{(for ideal gas)}$$

$$dU = 0 = Q - W \rightarrow W = Q \quad \text{(for ideal gas)}$$

Isochoric Process of the Ideal Gas

The isochoric process can be expressed with the ideal gas law as:

$$\frac{p}{T} = \text{constant}$$

Or

$$\frac{p_1}{T_1} = \frac{p_2}{T_2}$$

On a p-V diagram, the process occurs along a horizontal line that has the equation V = constant.

Table: Isochoric process – main characteristics.

First Law	$dU = dQ$
Ideal Gas Relation	$\frac{p}{T} = \text{constant}$
p, V, T Relations	$\frac{p_1}{T_1} = \frac{p_2}{T_2}$
Change in Internal Energy	$dU = mc_V(T_2 - T_1)$
Heat Transfer	$dQ = mc_V(T_2 - T_1)$
pdV Work	$W_{i \rightarrow f} = 0$

Isobaric Process

An isobaric process is a thermodynamic process, in which the pressure of the system remains constant (p = const). The heat transfer into or out of the system does work, but also changes the internal energy of the system. Since there are changes in internal energy (dU) and changes in system volume (ΔV), engineers often use the enthalpy of the system, which is defined as:

$$H = U + pV$$

Isobaric Process and the First Law

The classical form of the first law of thermodynamics is the following equation:

$$dU = dQ - dW$$

In this equation dW is equal to dW = pdV and is known as the boundary work. In an isobaric process and the ideal gas, part of heat added to the system will be used to do work and part of heat added will increase the internal energy (increase the temperature). Therefore it is convenient to use the enthalpy instead of the internal energy.

Isobaric process (Vdp = 0)

$$dH = dQ \rightarrow Q = H_2 - H_1$$

At constant entropy, i.e. in isentropic process, the enthalpy change equals the flow process work done on or by the system.

Isobaric Process of the Ideal Gas

The isobaric process can be expressed with the ideal gas law as:

$$\frac{V}{T} = \text{constant}$$

or

$$\frac{V_1}{T_1} = \frac{V_2}{T_2}$$

On a p-V diagram, the process occurs along a horizontal line (called an isobar) that has the equation p = constant.

Table: Isobaric process – main characteristics.

First Law	$dU = dQ - pDV$ $dH = dQ - Vdp = dQ$
Ideal Gas Relation	$\frac{V}{T} = \text{constant}$
p, V, T Relations	$\frac{V_1}{T_1} = \frac{V_2}{T_2}$
Change in Internal Energy	$dU = mc_V(T_2 - T_1)$
Change in Enthalpy	$dH = mc_p(T_2 - T_1)$
Heat Transfer	$dQ = mc_p(T_2 - T_1)$

pdV Work	$W_{i\to f} = p(V_f - V_i)$ $W_{i\to f} = nR(T_f - T_i)$
Vdp Work	$W_{i\to f} = 0$

Diesel Cycle – pV, Ts Diagram

Diesel cycles are often plotted on a pressure-volume diagram (pV diagram) and on a temperature-entropy diagram (Ts diagram).

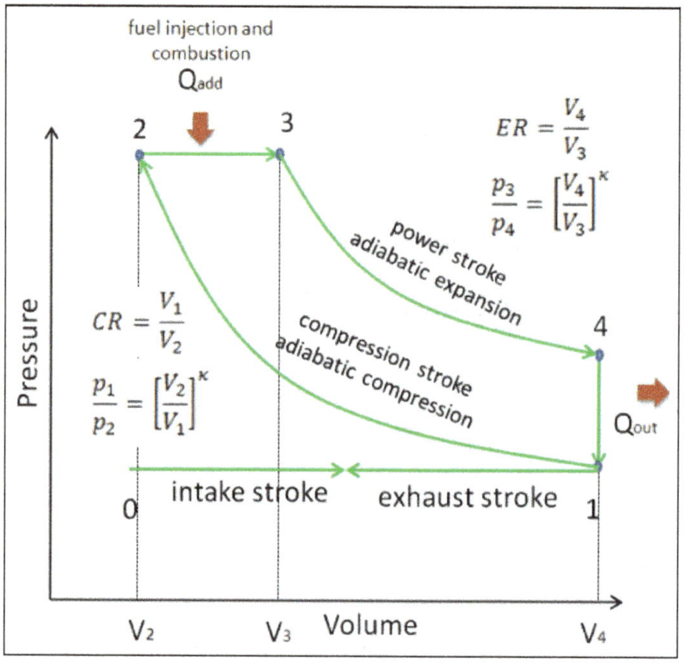

pV diagram of an ideal Diesel cycle.

When plotted on a pressure volume diagram, the isobaric process follow the isobar line for the gas (the horizontal lines), the isochoric process follow the isochoric line for the gas (the vertical line), adiabatic processes move between these lines and the area bounded by the complete cycle path represents the total work that can be done during one cycle. The temperature-entropy diagram (Ts diagram) in which the thermodynamic state is specified by a point on a graph with specific entropy (s) as the horizontal axis and absolute temperature (T) as the vertical axis. Ts diagrams are a useful and common tool, particularly because it helps to visualize the heat transfer during a process. For reversible (ideal) processes, the area under the T-s curve of a process is the heat transferred to the system during that process.

Four Stroke Diesel Engine

Diesel engines may be designed as either two-stroke or four-stroke cycles. The four-stroke Diesel engine is an internal combustion (IC) engine in which the piston completes four separate strokes while turning a crankshaft. A stroke refers to the full travel of the piston along the cylinder, in either direction.

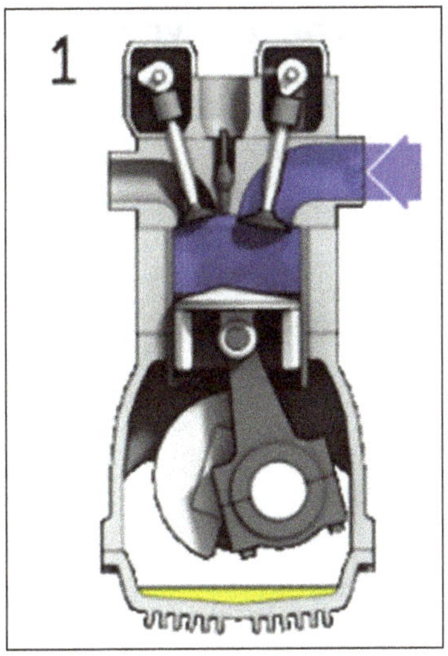

The Diesel engine is similar in operation to the gasoline engine. In this picture, there is an Otto engine, which is ignited by a spark plug instead of compression itself. Four stroke engine – Otto engine.

The four-stroke engine comprises:

- The Intake Stroke: The piston moves from top dead center (TDC) to bottom dead center (BDC) and the cycle passes points 0 → 1. In this stroke the intake valve is open while the piston pulls air (without a fuel) into the cylinder by producing vacuum pressure into the cylinder through its downward motion.
- The Compression Stroke: The piston moves from bottom dead center (BDC) to top dead center (TDC) and the cycle passes points 1 → 2. In this stroke both the intake and exhaust valves are closed, resulting in adiabatic air compression (i.e. without heat transfer to or from the environment). During this compression, the volume is reduced, the pressure and temperature both rise. At the end of this stroke fuel is injected and burns in the compressed hot air. At the end of this stroke the crankshaft has completed a full 360 degree revolution.
- The Power Stroke: The piston moves from top dead center (TDC) to bottom dead center (BDC) and the cycle passes points 2 → 3 → 4. In this stroke both the intake and exhaust valves are closed. At the beginning of the power stroke, a near isobaric combustion occur between 2 and 3. In this interval the pressure remains constant since the piston descends, and the volume increases. At 3 fuel injection and combustion are complete, and the cylinder contains gas at a higher temperature than at 2. Between 3 and 4 this hot gas expands, again approximately adiabatically. In this stroke the piston is driven towards the crankshaft, the volume in increased, and the work is done by the gas on the piston.
- The Exhaust Stroke: The piston moves from bottom dead center (BDC) to top dead center (TDC) and the cycle passes points 4 → 1 → 0. In this stroke the exhaust valve is open while the piston pulls an exhaust gases out of the chamber. At the end of this stroke the crankshaft has completed a second full 360 degree revolution.

In an ideal case, the adiabatic expansion should continue, until the pressure falls to that of the surrounding air. This would increase the thermal efficiency of such engine, but this also causes the practical difficulties with the engine. Simply the engine would have to be much larger.

Comparison of Actual and Ideal Diesel Cycles

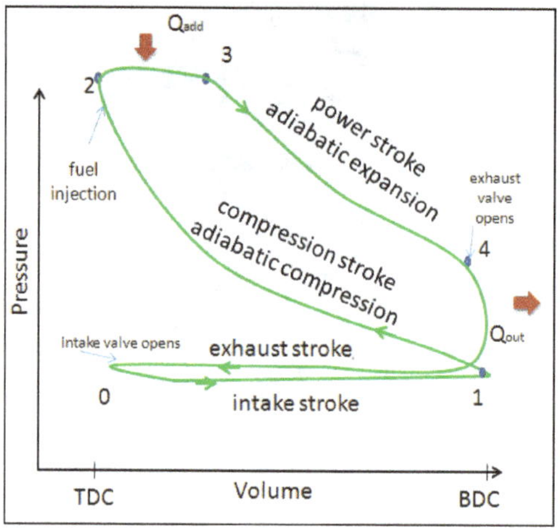

It is shown an ideal Diesel cycle in which there are a lot of assumptions differs from actual Diesel cycle. The main differences between the actual and ideal Diesel engine appear in the figure. In reality, the ideal cycle does not occur and there are many losses associated with each process. For an actual cycle, the shape of the pV diagram is similar to the ideal, but the area (work) enclosed by the pV diagram is always less than the ideal value. The ideal Diesel cycle is based on the following assumptions:

- Closed Cycle: The largest difference between the two diagrams is the simplification of the intake and exhaust strokes in the ideal cycle. In the exhaust stroke, heat Q_{out} is ejected to the environment (in a real engine, the gas leaves the engine and is replaced by a new mixture of air and fuel).
- Isobaric Heat Addition: In real engines the heat addition is never isobaric.
- No Heat Transfer:
    - Compression: The gas is compressed adiabatically from state 1 to state 2. In real engines, there are always some inefficiencies that reduce the thermal efficiency.
    - Expansion: The gas expands adiabatically from state 3 to state 4.
- Complete combustion of the mixture.
- No Pumping Work: Pumping work is the difference between the work done during exhaust stroke and the work done during intake stroke. In real cycles, there is a pressure difference between exhaust and inlet pressures.
- No Blowdown Loss: Blowdown loss is caused by the early opening of exhaust valves. This results in a loss of work output during expansion stroke.
- No Blow-by Loss: The blow-by loss is caused by the leakage of compressed gases through piston rings and other crevices.
- No frictional losses.

These simplifying assumptions and losses lead to the fact that the enclosed area (work) of the pV diagram for an actual engine is significantly smaller than the size of the area (work) enclosed by the pV diagram of the ideal cycle. In other words, the ideal engine cycle will overestimate the net work and, if the engines run at the same speed, greater power produced by the actual engine by around 20% (similarly as in the case of Otto engine).

Thermal Efficiency for Diesel Cycle

In general the thermal efficiency, η_{th}, of any heat engine is defined as the ratio of the work it does, W, to the heat input at the high temperature, Q_H.

$$\eta_{th} = \frac{W}{Q_H}$$

The thermal efficiency, η_{th}, represents the fraction of heat, Q_H, that is converted to work. Since energy is conserved according to the first law of thermodynamics and energy cannot be be converted to work completely, the heat input, Q_H, must equal the work done, W, plus the heat that must be dissipated as waste heat Q_C into the environment. Therefore we can rewrite the formula for thermal efficiency as:

$$\eta_{th} = \frac{W}{Q_H} = \frac{Q_H - Q_C}{Q_H} = 1 - \frac{Q_C}{Q_H}$$

The heat absorbed occurs during combustion of fuel-air mixture, when the spark occurs, roughly at constant volume. Since during an isochoric process there is no work done by or on the system, the first law of thermodynamics dictates $\Delta U = \Delta Q$. Therefore the heat added and rejected are given by:

$$Q_{add} = mc_p (T_3 - T_2)$$

$$Q_{out} = mc_v (T_4 - T_1)$$

Substituting these expressions for the heat added and rejected in the expression for thermal efficiency yields:

$$\eta_{th} = 1 - \frac{1}{\kappa}\frac{T_4 - T_1}{T_3 - T_2}$$

This equation can be rearranged to the form with the compression ratio and the cut-off ratio:

$$\eta_{Diesel} = 1 - \frac{1}{CR^{\kappa-1}}\left(\frac{\alpha^{\kappa} - 1}{\kappa(\alpha - 1)}\right)$$

Where,

- η_{Diesel} is the maximum thermal efficiency of a Diesel cycle.
- α is the cut-off ration V_3 / V_2 (i.e. the ratio of volumes at the end and start of the combustion phase).
- CR is the compression ratio.

- $\kappa = c_p / c_v = 1.4$

It is very useful conclusion, because it is desirable to achieve a high compression ratio to extract more mechanical energy from a given mass of the fuel. The air-standard Otto cycle thermal efficiency is also a function of compression ratio and κ.

$$\eta_{th} = 1 - \frac{T_4 - T_1}{T_3 - T_2} \rightarrow \eta_{otto} = 1 - \left(\frac{V_2}{V_1}\right)^{\kappa-1} = 1 - \frac{1}{CR^{\kappa-1}}$$

When we compare these to formulae, it can be seen that for a given compression ratio (CR), the Otto cycle will be more efficient than the Diesel cycle . But diesel engines are usually be more efficient since they are able to operate at higher compression ratios. In ordinary Otto engines the compression ratio has its limits. The compression ratio in a gasoline-powered engine will usually not be much higher than 10:1. Higher compression ratios will make gasoline engines subject to engine knocking, caused by autoignition an unburned mixture, if lower octane-rated fuel is used. In diesel engines there is minimal risk of autoignition of the fuel, because diesel engines are compression-ignition engines and there is no fuel in the cylinder at the beginning of the compression stroke.

Dual Cycle Engine or Semi-diesel Cycle Engine

Dual Combustion Cycle is a combination of Otto cycle and Diesel cycle. It is sometimes called semi-diesel cycle, because semi-diesel engines work on this cycle. In this cycle, heat is absorbed partly at a constant volume and partly at a constant pressure.

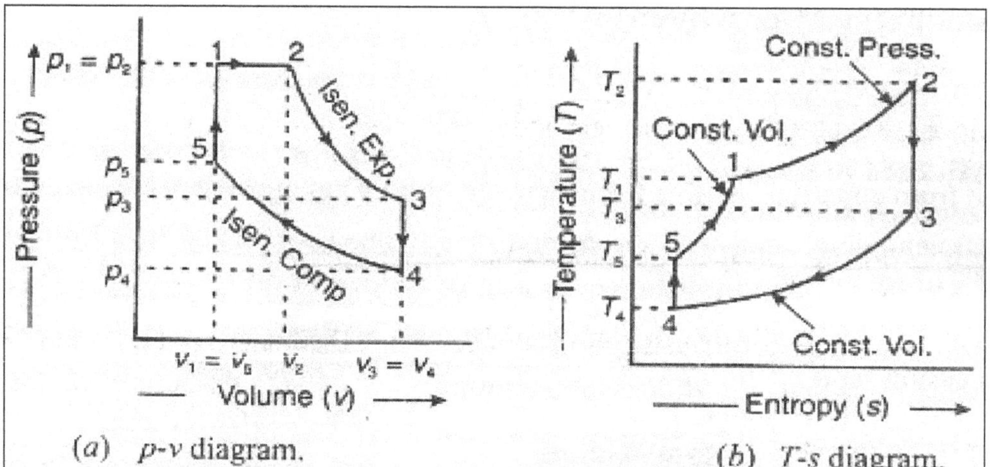

(a) p-v diagram. (b) T-s diagram.

The ideal dual combustion cycle consists of two reversible adiabatic or isentropic, two constant volume and a constant pressure processes. These processes are represented on p-v and T-s diagrams as shown in figure (a) and (b) respectively. The air standard efficiency of this cycle is given by:

$$\eta = 1 - \frac{T_3 - T_4}{\gamma(T_2 - T_1) + (T_1 - T_5)} = 1 - \frac{1}{r^{\gamma-1}}\left[\frac{\alpha.\rho^{\gamma} - 1}{(\alpha - 1) + \gamma.\alpha(\rho - 1)}\right]$$

$$r = \text{Compression ratio} = \frac{V_4}{V_5} = \frac{V_3}{V_1}$$

$$\rho = \text{Cut-off ratio } \frac{V_2}{V_1} = \frac{V_2}{V_5}$$

$$\alpha = \text{Pressure or expansion ratio } = \frac{p_1}{p_5}$$

According to Number of Strokes

Four-stroke Engine

A four-stroke cycle engine is an internal combustion engine that utilizes four distinct piston strokes (intake, compression, power, and exhaust) to complete one operating cycle. The piston make two complete passes in the cylinder to complete one operating cycle. An operating cycle requires two revolutions (720°) of the crankshaft. The four-stroke cycle engine is the most common type of small engine. A four-stroke cycle engine completes five Strokes in one operating cycle, including intake, compression, ignition, power, and exhaust Strokes.

Intake Stroke

The intake event is when the air-fuel mixture is introduced to fill the combustion chamber. The intake event occurs when the piston moves from TDC to BDC and the intake valve is open. The movement of the piston toward BDC creates a low pressure in the cylinder. Ambient atmospheric pressure forces the air-fuel mixture through the open intake valve into the cylinder to fill the low pressure area created by the piston movement. The cylinder continues to fill slightly past BDC as the air-fuel mixture continues to flow by its own inertia while the piston begins to change direction. The intake valve remains open a few degrees of crankshaft rotation after BDC. Depending on engine design. The intake valve then closes and the air-fuel mixture is sealed inside the cylinder.

Compression Stroke

The compression stroke is when the trapped air-fuel mixture is compressed inside the cylinder. The combustion chamber is sealed to form the charge. The charge is the volume of compressed air-fuel mixture trapped inside the combustion chamber ready for ignition. Compressing the air-fuel mixture allows more energy to be released when the charge is ignited. Intake and exhaust valves must be closed to ensure that the cylinder is sealed to provide compression. Compression is the process of reducing or squeezing a charge from a large volume to a smaller volume in the combustion chamber. The flywheel helps to maintain the momentum necessary to compress the charge. When the piston of an engine compresses the charge, an increase in compressive force supplied by work being done by the piston causes heat to be generated. The compression and heating of the air-fuel vapor in the charge results in an increase in charge temperature and an increase in fuel vaporization. The increase in charge temperature occurs uniformly throughout the combustion chamber to produce faster combustion (fuel oxidation) after ignition.

The increase in fuel vaporization occurs as small droplets of fuel become vaporized more completely from the heat generated. The increased droplet surface area exposed to the ignition flame allows more complete burning of the charge in the combustion chamber. Only gasoline vapor ignites. An increase in droplet surface area allows gasoline to release more vapor rather than remaining a liquid. The more the charge vapor molecules are compressed, the more energy obtained from the combustion process. The energy needed to compress the charge is substantially less than the gain in force produced during the combustion process. For example, in a typical small engine, energy required to compress the charge is only one-fourth the amount of energy produced during combustion. The compression ratio of an engine is a comparison of the volume of the combustion chamber with the piston at BDC to the volume of the combustion chamber with the piston at TDC. This area, combined with the design and style of combustion chamber, determines the compression ratio. Gasoline engines commonly have a compression ratio ranging from 6:1 - 10:1. The higher the compression ratio, the more fuel-efficient the engine. A higher compression ratio normally provides a substantial gain in combustion pressure or force on the piston. However, higher compression ratios increase operator effort required to start the engine. Some small engines feature a system to relieve pressure during the compression stroke to reduce operator effort required when starting the engine.

Ignition Event

The ignition (combustion) event occurs when the charge is ignited and rapidly oxidized through a chemical reaction to release heat energy. Combustion is the rapid, oxidizing chemical reaction in which a fuel chemically combines with oxygen in the atmosphere and releases energy in the form of heat. Proper combustion involves a short but finite time to spread a flame throughout the combustion chamber. The spark at the spark plug initiates combustion at approximately 20° of crankshaft rotation before TDC (BTDC). The atmospheric oxygen and fuel vapor are consumed by a progressing flame front. A flame front is the boundary wall that separates the charge from the combustion by-products. The flame front progresses across the combustion chamber until the entire charge has burned.

Power Stroke

The power stroke is an engine operation Stroke in which hot expanding gases force the piston head away from the cylinder head. Piston force and subsequent motion are transferred through the connecting rod to apply torque to the crankshaft. The torque applied initiates crankshaft rotation. The amount of torque produced is determined by the pressure on the piston, the size of the piston, and the throw of the engine. During the power Stroke, both valves are closed.

Exhaust Stroke

The exhaust stroke occurs whenspent gases are expelled from the combustion chamber and released to the atmosphere. The exhaust stroke is the final stroke and occurs when the exhaust valve is open and the intake valve is closed. Piston movement evacuates exhaust gases to the atmosphere. As the piston reaches BDC during the power stroke combustion is complete and the cylinder is filled with exhaust gases. The exhaust valve opens, and inertia of the flywheel and other moving parts push the piston back to TDC, forcing the exhaust gases out through the open exhaust valve. At the end of the exhaust stroke, the piston is at TDC and one operating cycle has been completed.

Five-stroke Engine

The four-stroke cycle has expansion and compression ratios which are almost identical and in order to make an engine more efficient we require higher expansion ratios[2] and lower compression ratios so as to limit the stresses due to mechanical and thermal factors. In order to make a better and efficient IC engine these limitations should be overcome and this was made possible with the help of introduction of a new cycle or a fifth stroke, also called as the recycle[3] stroke in the existing four-stroke cycle. By this extra stroke a configuration has been achieved which has independent or different expansion and compression ratios. The higher expansion ratios[2] help in extracting more and more thermal energy to convert into mechanical energy and lower compression ratios provides a mammoth range of load pressures. Hence, the necessity to increase engine quality parameters and make it eco-friendly by reducing fuel consumptiont gave birth to a new engine, called a 'Five Stroke Internal Combustion Engines'.

The twin-valve turbocharged direct injectionu five-stroke internal combustion engine is the latest technological advancement in the field of the automobiles with an emphasis on increasing the overall vehicular efficiency. The efficiency of an IC engine is mainly linked with its expansion ratio which helps in converting more thermal energy into mechanical energy and in order to make improvements in efficiency of an engine the compression ratios cannot be decreased but the increase in the expansion ratios can boost the efficiency of an IC engine. This non-symmetric configuration of the cycles can avoid and overcome the geometrical limits. The creative and innovative concept of having dual expansion cycles has made possible to develop an engine with high efficiency and large power range extenders which gives lower heat emissions and harmful gases. The engine generates the power output by the help of five strokes which are injection, compression, power, extended expansion (recycle stroke), exhaust. The engine has two high pressuret (HP) cylinders and one low pressuret (LP) cylinder, the volume of the low pressure cylinder is almost double[1] of that of the high pressure cylinder. The two high pressure compact cylinders help in maintaining lower compression ratios and works on the principle of four-stroke cycle. While the low pressure cylinder helps in extracting more work by an additional expansion stroke by taking the burnt fuel mixture from the high pressure cylinders which results in higher expansion ratios and lower temperature of the exhaust emissions. Since, the engine exhales from only one cylinder and hence its exhaust system can be made simpler and can also create more efficient impulse charging.

Construction Details of the New Concept Engine

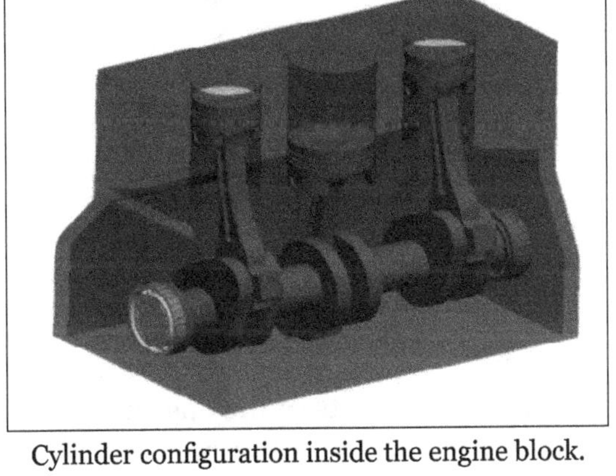

Cylinder configuration inside the engine block.

The engine consists of three cylinders which are mounted on a single crankshaft which rotates with a predetermined configuration required for the movement of the cylinders during the different strokes at different instants of time.

As shown in the figure the engine block has three cylinder configurations. The latter consists of two high pressure cylinders which works alternatively and have low volume when compared to the low pressure cylinder which has almost double volume as that of the high pressure ones. The low pressure cylinder is mounted in between the two high pressure cylinders which work in a cycle coordinating with the two high pressure cylinders and helps in providing more energy extraction by the help of an extra expansion stroke boosting the overall efficiency of the engine.

Exterior (casted) view of the engine block.

The figure shows the exterior view of the five-stroke engine block, made up of an alloy which is manufactured by casting process and has space for three cylinders which are fitted inside the chamber.

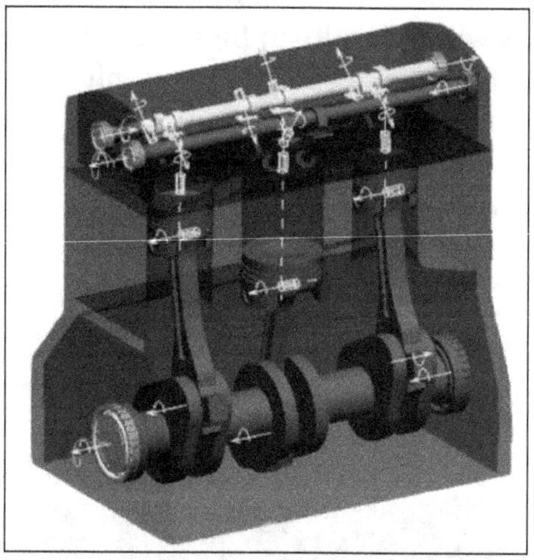

Inside view of the five-stroke engine block showing various components with the respective degrees of freedom.

The figure shows various important components of the engine assembly i.e., cylinders (HP-1, HP-2, LP), connecting rods, cams, valves, camshaft, crankshaft, etc. along with the respective degrees of freedom. It is clearly shown that the rotation of the shafts in required direction ultimately helps in the movement of the valves as well as the cylinders to work in the proper manner.

Working – Concept and Operating Principle

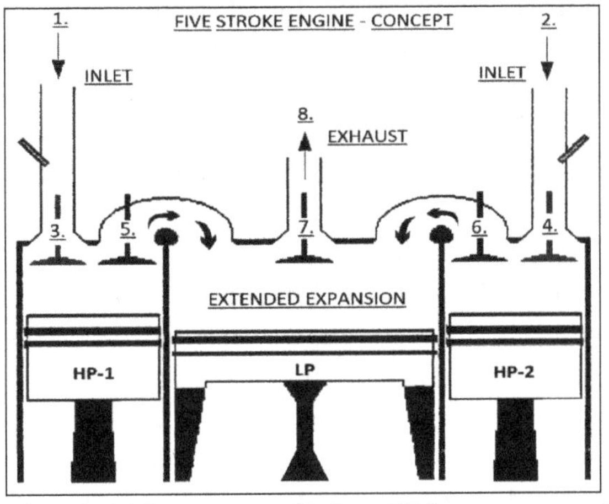

Five Stroke Engine Concept: HP-1/2-twin high pressure cylinders; LP-low pressure cylinder; 1, 2- intake channel; 3, 4- inlet valves for the two HP cylinders; 5, 6-exhaust valves into LP cylinder; 7-exhaust valve; 8-exhaust channel.

Table: The sequences of strokes producing work in five-stroke engine cylinders.

Stroke Seq.	Working Process in High Pressure Cylinder (HP-1)	Working Process in Low Pressure Cylinder (LP)	Working Process in High Pressure Cylinder (HP-2)
1.	Intake of air-fuel mixture.	Exhaust to outer environment.	Power generation.
2.	Compression of the mixture.	Extended expansion extracting more work.	Exhaust to LP cylinder.
3.	Power generation.	Exhaust to outer environment.	Intake of air-fuel mixture.
4.	Exhaust to LP cylinder.	Extended expansion extracting more work.	Compression of the mixture.
5.	Intake of air-fuel mixture.	Exhaust to outer environment.	Power generation.

As shown in figure the five-stroke engine works with the concept of two high pressure cylinders working in a four-stroke cycle and a low pressure cylinder in an alternate cycle to give the power output. The reduction in many components due to limiting the number of cylinders to three on the single shaft, the weight of the engine is reduced significantly and leads to a more compact yet efficient engine. The first stroke comprises of the direct injection of the air-fuel mixture into one of the high pressure cylinder [HP-1] through the inlet channel when the piston is initially at the TDC and the inlet valve is opened, the injection causes longitudinal piston movement and it further moves to the BDC and intake the fuel mixture and further expansion takes place. The piston is then again moved to the TDC to a predetermined distance in order to complete the compression stroke which helps in compressing the fuel mixture at the high pressure which is required for the effective burning of the fuel, both valves beingclosed. After completing the compression stroke, the spark plug ignites the gasoline mixture which is already at a very high pressure and this result into an explosion due to which the piston moves to the BDC and this stroke of burning of the fuel mixture is called as the power stroke.

When the mixture is burned, the piston movement takes place and it moves to TDC and the exhaust valve is opened in order to exhaust the burnt mixture. This burnt mixture is taken into a low pressure cylinder [LP] which extracts more heat from the vapors exhausted at very high temperatures from the high pressure cylinder [HP-1] by the help of another stroke which is called as an extended expansion stroke during which the exhaust valve is closed and after completing one cycle the gases are exhausted to the atmosphere through the exhaust channel. Another high pressure cylinder [HP-2] works in the similar manner but alternatively and the burnt mixture is exhausted to the low pressure cylinder [LP] in an alternative cycle. During the intake of the air-fuel mixture in the high pressure cylinder [HP-1], the process of power generation takes place in the second twin cylinder [HP-2] and the process continues vice versa. The low pressure cylinder [LP] consists of only two cycles which are extended expansion and then ultimate exhaust to the atmosphere. The table above describes the sequences of the various strokes (1),(2),(3),(4),(5) which helps in producing power output from the two cylinders [HP-1/2], [LP]. When there is an intake stroke taking place inside the first high pressure cylinder [HP-1], the second twin cylinder [HP-2] generates power through a power stroke and the low pressure cylinder is at the TDC exhausting the air-fuel mixture to the atmosphere. When there is compression process taking place inside first high pressure cylinder [HP-1], the second cylinder exhausts the burnt mixture into the low pressure cylinder [LP] which processes an extended expansion stroke in order to extract more amount of work from the mixture which is already burnt and is at a very high temperature range. The three cylinders work in the same order and hence, producing the continuous power output in an effective manner and the cycles continues vice versa.

This effective process helps in generating a power output of 40kW[1] as the upper limit in the working range between 3500-4000 rpm[1] of the engine which is made possible by the high expansion ratio. The peak torque of 160Nm and a horsepower of over 185 per liter can be achieved by a 700cc three cylinder five-stroke engine. It can deliver a brake power output of 32.5kW[1] at the working range of 4000rpm with a very low BSFC of 226g/kW-hr[1] with an overall efficiency of 36.1%[1]. The engine delivers a high power density and can deliver the same power output with the volume of 750cc as a 1200cc classical four-stroke internal combustion engine could deliver which makes high power densityu generation in more compact size and hence it results in the reduction of the overall weight of the engine which can be up to 20% lighter. The extended stroke also helps in lowering the exhaust temperature and further lower emissions of harmful gases increasing the thermal efficiency of the engine.

Technical Data

A typical 750cc 5-stroke engine can produce equivalent torque and power when compared to a 1200cc 4- stroke engine of order 110N-m and 46kW respectively at the speed of 4000rpm.

Table: Technical data (5-Stroke).

Parameter	High Pressure Cylinder (HP)	Low Pressure Cylinder (LP)
Number of Cylinders	2	1
Displacement (cc)	300(150+150)	450
Bore (mm)	60	83
Stroke (mm)	53	83
Compression Ratio	7:1	17,7:1
Number of intake/exhaust valves(diameter)(mm)	1/1 (21)	1/1 (45)

The technical specifications of the typical 5-stroke engine are shown in the table. The engine consisting of three cylinders has an overall volume of 750cc(150+150+450) and the LP and HP cylinders have corresponding bore and stroke dimensions as shown in the table.

Calculation Results

The engine was simulated in virtual environment to analyze the behavior at different load conditionsv between the range of 5% to 100% at the engine speed of 4000rpm.

Fuel Economy

The calculated results shows that the range of fuel economy of 5-S vs. 4-S varies from 37.8% (at 5% load) to 16% (at 100% load)v when the engine speed is of the order of 4000rpm.

Table: Fuel economy of the 5-troke vs. 4-stroke cycle at 4000 rpm v.

Load	Effective Torque(N-m)		Global Efficiency(%)		BSFC(gm/kW-hr)		Economy(%)
	4-Stroke	5-Stroke	4-Stroke	5-Stroke	4-Stroke	5-Stroke	5-S vs. 4-S
5	5.5	5.5	11.1	17.8	757.9	471.3	37.8
10	11.1	10.9	17.6	26.4	478.3	318.6	33.4
20	22.1	22.0	24.8	35.2	338.6	239.2	29.4
30	32.9	33.0	28.7	38.2	292.8	220.2	24.8
40	44.0	43.8	31.2	39.0	269.4	215.8	19.9
50	55.0	54.7	32.9	39.7	255.7	211.7	17.2
60	66.0	65.7	34.1	40.5	246.7	207.5	15.9
70	77.1	76.7	35.0	41.4	240.3	203.0	15.5
80	88.1	87.9	35.7	42.3	235.7	198.9	15.6
90	99.1	98.8	36.2	43.1	232.3	195.2	16.0
100	110.0	109.9	36.6	43.6	229.6	192.9	16.0

The table above, gives detail information about the net effective torque(N-m), global efficiency, BSF-C(gm/kW-hr) and fuel economy at various load conditions of both 4-stroke as well as 5-stroke engine.

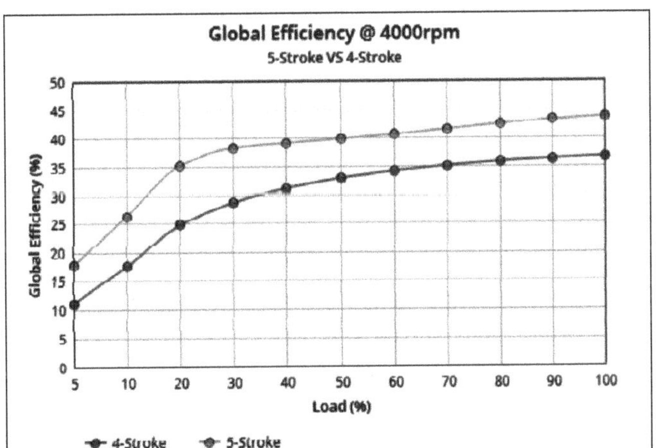

Global efficiency of 5-stroke vs. 4-stroke @4000rpm.

The figure shows the variation of the global efficiency of the 4-stroke cycle with the 5-stroke cycle at various load conditions when the engine is running at the speed of 4000rpm.

Figure below shows variation of the brake specific fuel consumption in gm per kilowatt hour at different load conditions of both 4-stroke and 5-stroke cycles when the engine speed is 4000rpm. The figure shows the fuel economy of 5-stroke cycle over 4-stroke cycle at the same speed which clearly shows that the 5- stroke cycle is much more efficient when compared to the typical 4-stroke cycle at same load and speed conditions.

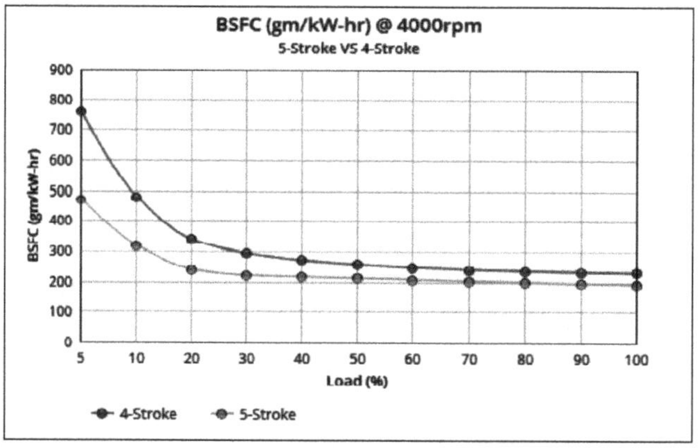

BSFC (gm/kW-hr) of 5-stroke vs. 4-stroke @4000rpm.

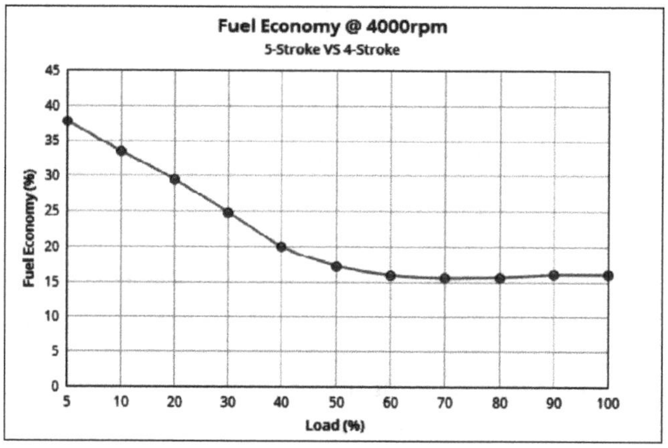

Fuel Economy 5S vs. 4S @4000rpm.

Six-stroke Engine

The six stroke engine is a type of IC engine based on the four Stroke engine but with additional complexity intended to make it more efficient and reduce emissions. Six stroke engine captures the wasted heat from the four-stroke Otto Cycle, Creates steam and cools engine by providing free Power Stroke developed by Bruce Crower and Malcolm Beare. The six stroke engine is a concept that has been around for over 100 years, but has only recently been reconsidered as a feasible alternative to the conventional Four Stroke Engine due to advancement in automotive and material technologies. Amidst the ever progressing automobile engineering comes a new trend of engines popularly known as the 6 stroke engines. The six stroke engines are mostly recommended for their extra power output and thus most preferred in heavy vehicles where the prime focus is on load carrying capacity rather than fuel economy. It consists of six strokes which are due to the radical hybridization of 2 strokes and 4 strokes, i.e. the piston in each move goes up and down six times for the injection of fuel. The six stroke engine consists of 2 chambers having internal combustion and external combustion wherein, the unused or waste heat from the 4 stroke Otto cycle is then used to carry out further strokes. These 2 additional strokes increase the work extracted per unit input of energy, which then lead to 40% increase in fuel efficiency than the four-stroke Otto cycle. The six strokes of the six stroke cycles are as follows:

- Intake Stroke.
- Compression Stroke.
- Ignition Stroke.
- 1st Exhaust Stroke.
- Steam Expansion Stroke.
- 2nd Exhaust Stroke.

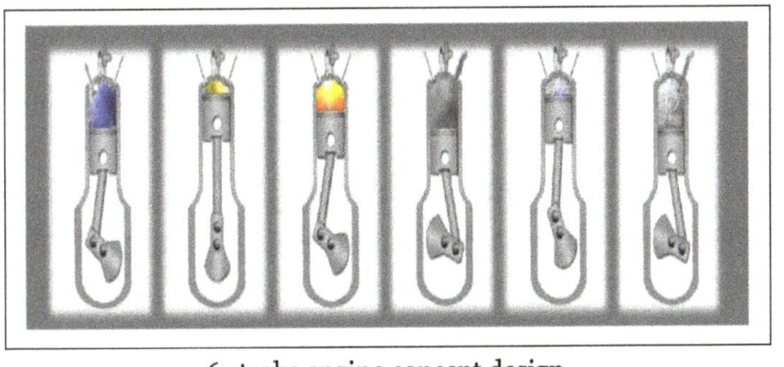

6 stroke engine concept design.

- Intake Stroke: The inlet valve open as the piston moves down from TDC to BDC which enters the air-fuel mixture into the chamber.
- Compression Stroke: The inlet valve close as the piston moves BDC to TDC and compress the air/fuel mixture under pressure.
- Ignition Stroke: The air/fuel mixture is then ignited, causing a small explosion which forces the piston back down which turns the crank and provides the torque.
- Exhaust Stroke: In The exhaust stroke the exhaust valve open as the piston moves back up once again , pushing the by products of the fuel explosion out of the chamber out of the chamber.
- Steam Expansion Stroke: After the exhaust stroke in six stroke Engine the water inside the Extremely hot chamber the water Immediately turns to steam – expanding To 1600 times its volume –which Force the piston down for a Second power stroke.
- 2nd Exhaust Stroke: In this stroke all the remaining vapors And gases are thrown away from the Chamber water vapors are collected by A condenser which is attached with the Exhaust port so that water can be used again.

Crankshaft

In four-stroke engine, the crankshaft rotates about 720 degree while the camshaft rotates 360 degree to complete one cycle. But in six stroke engine the crankshaft must rotate 1080 degree to rotate the camshaft 360 degree to complete one cycle. Thus the gear ratio of six stroke will be 3:1 instead 2:1 which is of four-stroke engine. The teeth of gear in crankshaft will be 18 and in camshaft the gear teeth will be 54.The gear used in this is helical type gear which gives high speed and high speed rotation.

OLD NEW

Camshaft

When camshaft rotates 360 degree in six stroke engine, the cam has been divided into 60 degree among the sixstrokes. There are two exhaust stroke, one will be at the time of fourth stroke through which hot gases or burnt gases are left out and the second one is at the six stroke which pushes the steam out.

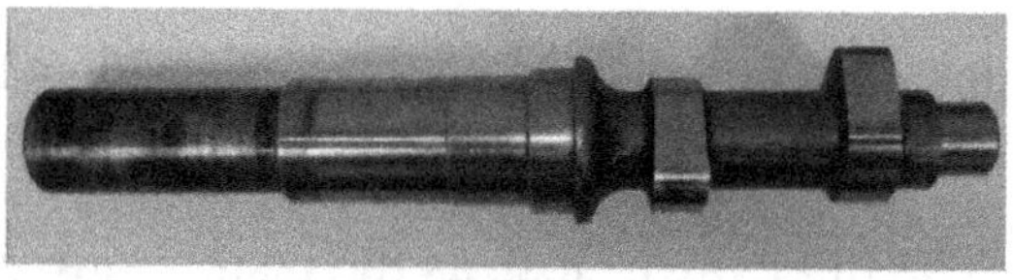

OLD

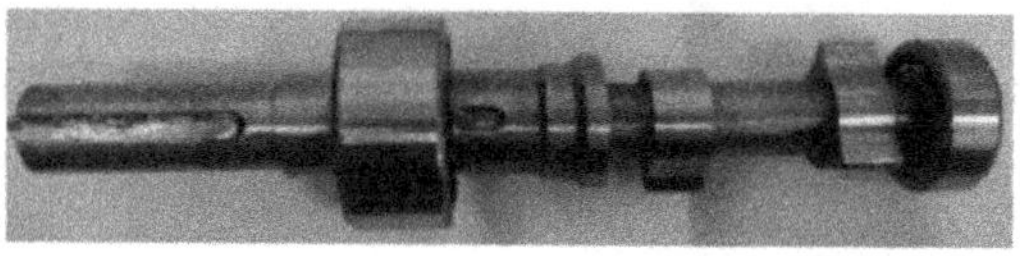

NEW

Advantage

- Thermal efficiency reaching 50% (30% for the actual internal combustion engine.
- Fuel consumption reduced by more than 40%.
- Reduction of chemical, noise and thermal pollution.
- Two expansion (work) through six stroke.
- Direct Injection and optimal fuel combustion at every engine speed.

- We can eliminate the radiator.
- We would get the better scavenging in fifth stroke.

Disadvantage

- High initial cost Due to change in gear structure.
- High manufacturing cost in six stroke engine.
- Engine size increases due to additional stroke in engine, fifth stroke and sixth stroke.

Two-stroke Engine

A two-stroke engine is a type of internal combustion engine that completes a power cycle with two strokes of the piston during only one crankshaft revolution. This is in contrast to a four-stroke engine that requires four strokes of the piston to complete a power cycle during two crankshaft revolutions.

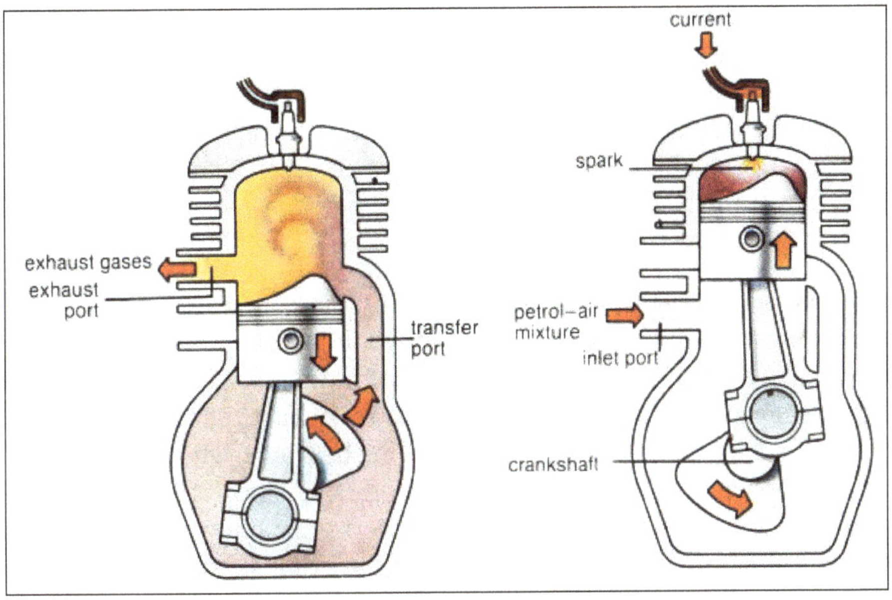

In a four-stroke engine, each of the four essential steps of the power-producing cycle is given its own piston stroke:

- Compression
- Power
- Exhaust
- Intake

A two-stroke engine performs all the same steps, but in just two piston strokes. The simplest two-stroke engines do this by using the crankcase and the underside of the moving piston as a fresh charge pump. Such engines carry the official name "crankcase-scavenged two-strokes." As the two-stroke's piston rises on compression, its underside pulls a partial vacuum in the crankcase. An intake port of some kind (cylinder wall port, reed valve or rotary disc valve) opens, allowing air to rush into the crankcase through a carburetor. As the piston nears Top Dead Center, a spark fires the compressed mixture. As in a four-stroke, the mixture burns and its chemical energy becomes heat energy, raising the pressure of the burned mixture to hundreds of psi. This pressure drives the piston down the bore, rotating the crankshaft.

As the piston continues down the bore, it begins to expose an exhaust port in the cylinder wall. As spent combustion gas rushes out through this port, the descending piston is simultaneously compressing the fuel-air mixture trapped beneath it in the crankcase. As the piston descends more, it begins to expose two or more fresh-charge ports, which are connected to the crankcase by short ducts. As pressure in the cylinder is now low and pressure in the crankcase higher, fresh charge from the crankcase rushes into the cylinder through the fresh-charge (or "transfer") ports. These ports are shaped and aimed to minimize direct loss of fresh charge to the exhaust port. Even in the best designs, there is some loss, but simplicity has its price. This process of filling the cylinder while also pushing leftover exhaust gas out the exhaust port is called "scavenging." While the piston is near Bottom Dead Center, mixture continues to move from the crankcase, up through the transfer ports, and into the cylinder. As the piston rises, it first covers the transfer ports, leaving only the exhaust port still open. If there were no way to stop it, much of the fresh charge would now be pumped out the exhaust. But there is a simple way to stop it—using exhaust pressure waves in the exhaust. If we shape and dimension the exhaust pipe right, a reflection of the original pressure pulse, generated as the exhaust port opened, will bounce back to the port just as fresh charge is being pumped out of it. This pressure wave stuffs the fresh charge back into the cylinder just as the rising piston covers the exhaust port.

Because fuel-air mixture is constantly being pumped by the crankcase, it is not practical to lubricate piston and crank by pumped circulating oil—it would be swept away by the mixture rushing in and out. Therefore, we must either mix a little oil with the fuel (2 to 4 percent) or inject it very sparingly into the bearings with a tiny metering pump. The fact that there is so little oil dictates that such simple two-stroke engines must employ rolling bearings, whose need for oil is very small. More complicated two-stroke engines exist. Instead of using the crankcase and underside of the piston as a fresh-charge pump, we can use a separate rotary blower, directly connected to the transfer ports in the cylinders. We don't have to place the exhaust port in the cylinder wall—it can take the form of four overhead poppet exhaust valves, as it does in two-stroke marine, rail, and truck diesels. Because such engines do not use their crankcases as fresh charge pumps, they can employ long-lasting plain bearings, lubricated conventionally by pumped recirculating oil. Two-stroke diesels are scavenged with pure air, not a fuel-air mixture. Their fuel is injected only after all ports have closed, preventing any loss. Certain crankcase scavenged two-strokes do the same, and are called "DI," or Direct Injection two-strokes. They can be made as fuel-efficient and low in exhaust emissions as four-strokes.

Two-stroke Engine Cycle

Down Stroke

The piston moves from TDC (Top-Dead-Center) to BDC (Bottom-Dead-Center) letting the fresh air enter into the combustion chamber. The fresh air-fuel mixture gets into the combustion chamber through the crankcase. In this stroke, the crankshaft makes the rotation of 180°.

Up Stroke

The piston is pushed from BDC to TDC. As a result, the fuel-air mixture gets compressed and the spark plug ignites the mixture. The mixture expands and the piston is pushed down. The inlet port is open during the upstroke. While the inlet port is opened, the mixture gets sucked inside the crank-case. When the mixture is pushed up into the combustion chamber during the previous upstroke, a partial vacuum is created as no mixture is left behind in the crankcase. This mixture is ready to go into the combustion chamber during downstroke but remains in the crankcase until the piston goes up till TDC. In this stroke, the crankshaft makes the rotation of 180°. From the 2nd downstroke onwards the exhaust gases get expelled out from one side while a fresh mixture enters into the combustion chamber simultaneously due to partial vacuum created in the combustion chamber after removal of exhaust gases. This is the beauty of the engine. Both things happen at the same time which makes it a 2-stroke engine. The exhaust gases are expelled from the 2nd downstroke onwards from one side while simultaneously a fresh mixture of air and fuel is injected into the combustion chamber due to the partial vacuum created in the combustion chamber after the removal of exhaust gases.

Applications of Two-stroke Engine

- Two-stroke engines are preferred when mechanical simplicity, lightweight, and high power-to-weight ratio are design priorities.
- They are lubricated by the traditional method of mixing oil into the fuel, they can be worked within any orientation as they do not have a reservoir dependent on gravity. This makes them desirable for their use in handheld tools such as chainsaws.
- Two-stroke engines are found in small scales propulsion applications such as motorcycles, mopeds, and dirt bikes.

According to Number of Cylinders

Single Cylinder Engine

A single cylinder engine is a type of combustion engine that features only one cylinder, or chamber in which a piston moves to engage combustion. This is the engine's source of power, and since only one piston is doing the work, a single cylinder engine is primarily used on smaller vehicles and tools. It can produce a relatively large amount of power given its size, but this engine is not especially adept at varying the power output quickly, making it less versatile for larger vehicles. These engines are lightweight and compact, making them a good choice for engine powered tools such as weed whackers.

Disadvantages of a single cylinder engine include the vibration the engine is known for producing, as well as the lower power output. Vibration occurs for several reasons; primary among them is the weight of the rotating parts. Since only one cylinder is used for power output, many of the rotating components have to be extra strong to withstand the strain of operation. This means the parts can be quite heavy — much heavier than their counterparts in two- and four-cylinder engines. This rotational weight can cause vibration as the piston activates, which is not too much of a disadvantage on small engines used for weed whackers, chain saws, and so on. It can be a problem on this type of engine mounted to a motorcycle or moped.

One of the biggest advantages of a single cylinder engine is the production cost. Since the engine is relatively small and does not feature several combustion chambers, the engine can be built cheaply and easily, driving down the cost of this effective engine style. These engines are therefore common on lower-end small motorcycles, mopeds, go-karts, and lawnmowers. The engine is also lightweight in most cases, making it easy to manipulate on handheld garden or yard machinery. Early motorcycle models used single cylinder engine designs as a power source, though the vibration issues made the engine a poor choice, especially in comparison to two- and four-cylinder engines. To counteract the vibration problems, some engines were fitted with balances that sought to steady the vibrating engines, and these worked well, to a point. As motorcycles grew larger and faster, however, the single-cylinders were replaced with larger engines that could power the machine more effectively. Automobiles have also used single cylinder engines in the past, but four-, six-, and eight-cylinder engines are far more common.

Double Cylinder Engine

The double cylinder engine constructed with the two cylinders that can be in same engine block or in different block. In double cylinder motorcycle engine both the cylinder performs separate combustion process. But the pistons of both cylinder stay connected with same crankshaft and delivers the power to the same. Hence it's significantly efficient of producing and delivering the power as there is very less power delivery gap to rotate the crankshaft than the single cylinder engine. According to performance double cylinder engine is much efficient of producing and delivering the power. In addition it's also much faster and sharper near like the single cylinder engine. For these special characteristics the light weighted but durable performance bikes are developed with double cylinder engines. Hence the legendary V-Twin, Parallel Twin, Flat-Twin engines are so popular in the motorcycle world.

Advantages of Double Cylinder Motorcycle Engine

- Double cylinder engines quite have both the good qualities of single cylinder and multiple cylinder engines.
- It's much more durable and efficient than the single cylinder engine.
- This type of engine is also much crisp, fast responding and sharp in character like the single cylinder engine.
- The engine has much less vibration than the single cylinder engine. Moreover offset finish of cylinder and crankshaft allows smoother engine operation in double cylinder engines.
- To keep the crankshaft rotation précised it doesn't need heavy balance shaft like the single cylinder engine.
- The features like traction control, engine braking etc. is much more efficient in double cylinder motorcycle engines.

Disadvantages of Double Cylinder Motorcycle Engine

- Double cylinder engine is little heavier than the single cylinder engine and its takes much more space.

- Its construction is not so complex like the multiple cylinder engine but comparatively complex than the single cylinder one.
- These engines are not high maintenance engine but its maintenance is little complex.
- The electricals are quite complex in double cylinder engine motorcycles and are sensitive.
- Liquid cooling system is the best solution for double cylinder engine but air cooling is possible though it's not the best solution.
- This type of engine is little less fuel efficient than the single cylinder engine. But that is really not so significant in modern designed double cylinder engines.
- Double cylinder engines are heavier than the single cylinder engine motorcycle. Hence for commuters and hardcore off road bikes this type of engine is not best option.

Multi-cylinder Engine

A multi-cylinder engine is a reciprocating internal combustion engine with multiple cylinders. The cylinders and the crankshaft which is driven by and co-ordinates the motion of the pistons can be configured in a wide variety of ways. Multi-cylinder engines offer a number of advantages over single-cylinder engines, chiefly with their ability to neutralize imbalances by having corresponding mechanisms moving in opposing.

Multi-cylinder Cyclic-torque

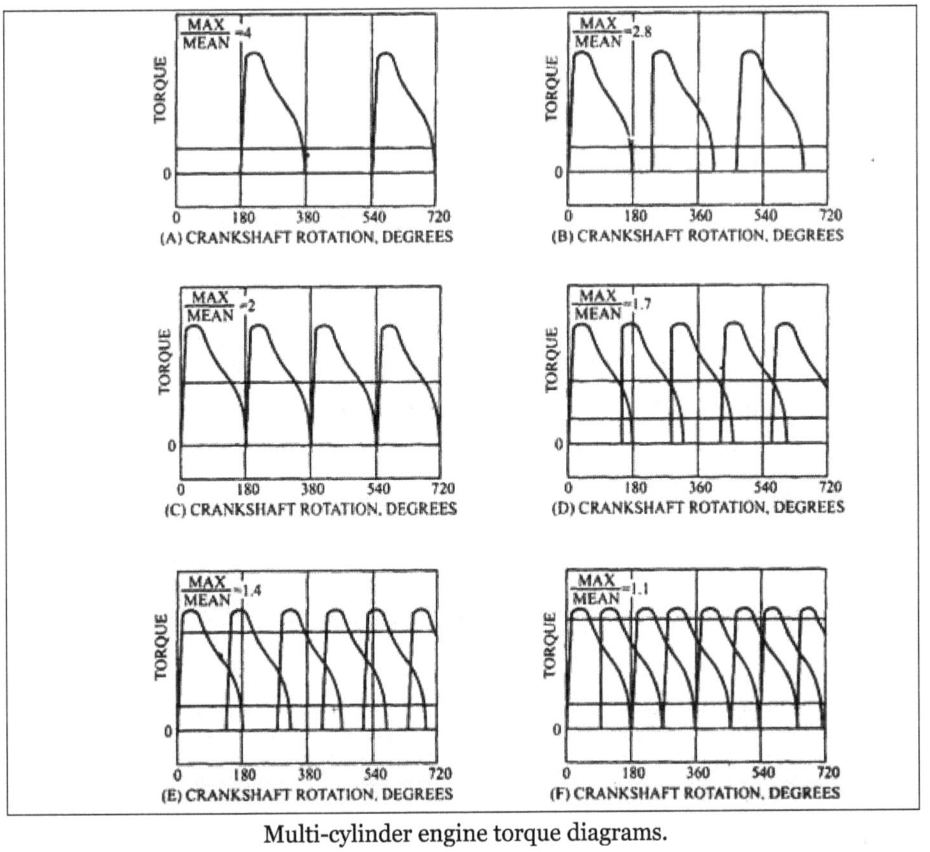

Multi-cylinder engine torque diagrams.

In figure, A. Twin-cylinder engine B. Three-cylinder engine. C. Four-cylinder engine D. Five-cylinder engine. E. Six-cylinder engine F. Eight-cylinder engine.

The limitations on the size of the flywheel and its inability to smooth out the torque unevenness between the cycles has been largely resolved by using multi-cylinder engines where the valve timing with the single crankshaft is sequenced so that the power strokes of the cylinders occur in a phased manner, instead to take place all at the same time. When the number of cylinders is increased, accordingly the intervals between power impulses reduce. Consequently, the torque variation throughout the four strokes of the cycle is smoothed out. The cyclic-torque curve for a single-cylinder engine shows a firing stroke every 720 degrees and the variation of peak to mean torque over one cycle is around 8:1. When a second cylinder is added, the interval between the firing impulses is halved, i.e., 360 degrees, thereby reducing the peak to mean torque produced over a cycle to 4:1. By adding a third cylinder, the interval between the firing impulses reduces to 240 degrees, and the peak to mean turning-effort is further smoothed out to the order of 2.8:1. The four-, five-, six-, and eight-cylinder engines have firing intervals of 180 degrees, 144 degrees, 120 degrees, and 90 degrees respectively, with respective ratios of peak to mean torque are reduced to 2:1, 1.7:1, 1.4:1, and 1.1:1(Figure C through F).

Merits and Limitations of Single- and Multi-cylinder Engines

The following major factors are required to be considered while comparing engines of different cubic capacity and various numbers of cylinders:

- For a given maximum piston speed the shorter the piston stroke, the higher can be the crankshaft rotation.
- As the cylinder becomes smaller, the piston becomes lighter in proportion to the cylinder size, accordingly causing higher piston speeds.
- For the same engine cylinder capacity and maximum piston speed, a multi-cylinder engine develops more power than a single-cylinder engine.
- A single-cylinder engine with the same piston cross-sectional area as a multi-cylinder engine produces a greater torque output.
- The smaller the cylinder size, the higher is its surface-to-volume ratio and hence higher is the compression-ratio with an improvement in engine thermal efficiency.
- For a given total volume, acceleration response improves with the number of cylinders, because of lighter reciprocating components and the smaller flywheel.
- As the number of cylinders and the engine length increase torsional vibration becomes a problem.
- As the number of cylinders increases:
    - The power consumed in overcoming rotational and reciprocating drag also increases.
    - Mixture distribution for carburetted engines becomes more difficult.

- The cost of replacement of components becomes proportionally higher.
- The frequency of power impulses increases, due to which the power output becomes more consistent.

Smooth operation of multi-cylinder engines is possible only when each combustion chamber produces the same combustion chamber pressure as others in the same engine. The carburettor should ensure the charge quality by mixing the fuel into the incoming air in the correct proportions. The intake manifold should direct an equal quantity of the mixed charge to each intake valve. Each intake valve must be timed same as the others to allow an equal quantity of the charge to enter each combustion chamber. The ignition distributor must be timed to send a spark across the spark plug gap when compression is attained to the same amount in all the cylinders. When all these requirements are met the pressure in the combustion chambers is equal. But practically these ideal requirements are not met under all operating conditions due 'to an increase in manufacturing cost. Multi-cylinder engines are preferred to single-cylinder engines, which would give the same output due to the following reasons:

Table: Difference between Large single-cylinder engine and Multi-cylinder engine.

Large single-cylinder engine	Multi-cylinder engine
• Jerky torque from only one power stroke per two revolutions. • Heavy flywheel required. • Large piston and valves present considerable cooling difficulties. • Large exhaust pulsations cause difficulty in silencing. • The engine would be very tall and difficult to accommodate under the bonnet Engine would be vary heavy. • The heavy piston poses difficulty in balancing. • Must run at low-speeds.	• More power strokes per revolution giving smooth torque output. • Lighter flywheel allowing quicker acceleration. • Small valves and pistons enable cooling easier. • More frequent and smaller pulsations make easier silencing. • Engine is much more compact. • Engine would weight much lesser than the single-cylinder engine. • Easy to balance. • Could be run at much higher speed.

According to Arrangement of Cylinders

Radial Engine

Radial engine is a type of internal-combustion engine used mainly in small airplanes, in which the cylinders (ranging from five to as many as 28, depending on engine size) are mounted in a circle around the crankshaft, sometimes in banks of two or more. Once the dominant piston-engine type, radials are now in only limited production; most new requirements are met by remanufacturing existing stock. A radial engine works like any other four-stroke internal combustion engine. Each cylinder has an intake, compression, power and exhaust stroke. They differ from inline and horizontally opposed engines in their firing order and they way they connect to the crankshaft. A radial engine's cylinders are numbered from the top, going clockwise, with the first cylinder numbered 1. The first cylinder's connecting rod attaches directly to the crankshaft-this is the master rod. The other cylinders' rods connect to pivoting points around the master rod.

Every radial engine has an odd number of cylinders, and they fire in an alternating order. So, a five-cylinder engine fires in a 1, 3, 5, 2, and 4 order. A seven-cylinder engine fires in an 1, 3, 5, 7, 2, 4, 6 order.

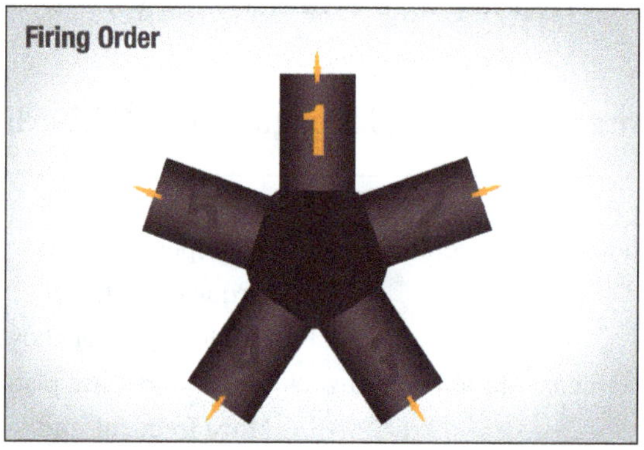

As the cylinders fire, the rod assembly rotates around the crankshaft, spinning it like a bell crank would. A counterbalance weight sits opposite from the rod hub to prevent engine vibration.

Advantages

- Easier Maintenance: The overall design of the radial engine makes it much easier for maintenance workers to access and provide routine maintenance work on the engine. In other words, it's easier to keep it running at its best.
- Tougher: Thanks to the simpler design, radial engines are usually less vulnerable to damage. While this was a much bigger deal when the engines were used in combat, knowing that an engine can stand up better than another option is still a good reason to consider it.
- Smoother Operation: In most cases, a smooth operation is easier to get with a radial engine than other options like inline engines cooled with liquid. Smooth performance matters, and the radial engine is generally a better call where performance is concerned.
- Reliability: Simply put, the radial engine is generally much more reliable. This is because it features a shorter crankshaft, simpler design, and creates less vibration. That less vibration means that it will suffer from greatly reduced levels of wear and tear during use.
- Versatility: Radial engines can be attached to a wide range of aircraft, and are among the most widely used and versatile options on the market today.

- Overall Cost: Almost everything about the radial engine is a bit less expensive, including purchasing price, repair costs, maintenance cost, and operating costs. In a field that is already very expensive to work in, a cheaper option is usually welcome.

Disadvantages

- Drag: As the cylinders are clearly exposed to the airflow during flight, drag increases considerably.
- Visiblity: The pilot may experience visibility problems, due to huge single engines located near the nose of aircrafts.
- Installation: It is more difficult to ensure adequate cooling air in a buried engine installation.

Applications

Radial engines have several advantages for airplanes:

- They can produce a lot of power. A typical radial engine in a B-17 has nine cylinders, displaces 1,800 cubic inches (29.5 liters) and produces 1,200 horsepower.
- Radial engines have a relatively low maximum rpm (rotations per minute) rate, so they can often drive propellers without any sort of reduction gearing.
- Because all of the pistons are in the same plane, they all get even cooling and normally can be air-cooled. That saves the weight of water-cooling.

The engine from a Harley can be thought of as two cylinders from a radial engine.

Radial engines reached their zenith during WWII. There are some radial engines around today, but they are not that common. Most propeller-driven planes today use more traditional engine configurations (like a flat four-cylinder) or modern gas turbine engines. Gas turbines are much lighter than radial engines for the power they produce. One place where you can still see the influence of the radial engine concept is in the two-cylinder engine of a Harley Davidson motorcycle.

It can be thought of, in a way, as two pistons from a radial engine. In a Harley, both pistons share a single connection point to the crankshaft, like in a radial engine.

V Engine

A V engine, also known as a vee engine, is a common type of internal combustion engine. It is characterized by the "V" shape that the cylinders make when viewed from the front of the engine, perpendicular to the axis of the crankshaft. The proponents of the V engine design often cite its generally shorter length, height and weight as reasons for its allure.

A V engine incorporates a single crankshaft.

The conventional notation used to designate the size of a V engine is the letter "V" followed by the number of cylinders the engine has. For instance, a V engine with six cylinders is called a V-6. A V configuration engine can be as small as a V-twin with only two cylinders mirroring each other across the crankshaft or as large as a V-24 in locomotives. The V engine consists of two banks, or rows, of cylinders that attach to a single crankshaft. Typically the cylinders are situated at a 60- or 90-degree angle. The banks in a V-6 are usually 60 or 90 degrees, and the banks of a V-8 engine usually lie in a 90-degree angle. This type of engine is commonly arranged into configurations of V-6, V-8, V-10 and V-12 in automobiles, though they are not exclusive to automobile usage. V-twins are a staple of some motorcycle manufacturers, and others have found ways to incorporate V engines ranging from a V-4 or V-5 to a V-8 in their bikes. V configuration engines V-16, V-20 and V-24 are quite uncommon in automobile use and are restricted to mainly large truck and locomotive usage, with some exceptions.

The appeal of the V engine design comes from its tendency to be lighter and shorter in height and length than an inline layout of the same cylinder number and displacement. The short length of the V design allows automotive producers to shorten the length of the hood and use that space for other purposes. The V engine tends to permit higher torsional stresses and rotational speeds as a result of its strong short crankshaft. The inline engine configuration, by contrast, is a much simpler design than the V, with a single bank containing all of the cylinders in a row. The larger block and distance between cylinders often cause the inline to have a smoother power delivery while producing more torque. Inline layouts are heavier and much longer than V-type engines, though, and they require more space in terms of length.

V4 Engines

A V4 engine is a a V form engine with four cylinders. It was first used by Lancia starting 1922 through 1960. It was a narrow-angle aluminum design. All three displacements shared the same long 120 mm stroke, and all were SOHC designs with a single camshaft serving both banks of cylinders.

Two totally diferent engines were produced by Ford of Europe: one was the V4 engine used in the british Ford Essex and the second one the V4 engine of the german Ford Taunus. The first one was available in two capacities: 1633cc and 1996cc, differing only in stroke. The second one was a V4 piston engine with one balance shaft, introduced by Ford Motor Company in Germany in 1962. Later it was later expanded into the Ford Cologne V6 engine that is used in the Ford Capri and many other Ford cars. The Taunus V4 was replaced by the Ford OHC/Pinto engine.

V6 Engine

A V6 is an internal combustion piston engine with six cylinders in a "V" configuration. It is the second most common engine configuration in modern cars after the inline-4; it shares with that engine a compactness very suited to the popular front wheel drive layout, and is becoming more popular as car weights increase.

The first V6 was introduced by Lancia on the 1924 Lancia Lambda, but it made little impact. The design was reintroduced by the company in 1950 with the Lancia Aurelia; this time, other manufacturers took note and soon other V6 engines were in use. The design really took off after the 1962 introduction of the Buick Special. Though the model was not a spectacular success, it was the first mass-produced V6 engine. In 1983 Nissan produced Japan's first V6 engine with the VG series.

V8 Engines

A V8 engine is a V engine with eight cylinders. The V8 is a very common configuration for large automobile engines. V8 engines are rarely less than 4 L in displacement and in automobile use have gone up to 8.5 L or so. The first engine was a british one, developed by Rolls Royce: a 3.5 L Rolls-Royce Legalimit. But Cadillac was the first automobile maker to mass produce a V8 engine. The company has produced eight generations of V8s since 1914. The Type 51 was the first Cadillac V8. Introduced in 1914, it was the standard engine for 1915 Cadillac models. One significant innovation with the 70-horsepower, 314-cubic-inch (5.1 liter) L-head design was the thermostatic control of cooling-water circulation. The engine, multi-plate clutch and gearbox were combined in one bolted-together assembly. The United States War Department purchased over 2,000 standard Cadillac V8 models for use in Europe during World War I.

The V-8 engine was developed by France's renowned auto maker, Count De Dion Bouton, so although it came as a distinct innovation in a stock American car, the principle was not a new one. Cadillac's new V-8 engine was actually shorter and 60 lbs lighter than the four-cylinder motor that preceded it. Its high torque allowed speeds up to 55-60mph to be achieved in high gear. Quick acceleration was another benefit of the new 314 cubic inch motor rated at 31.25HP (SAE); dynamometer tests produced 70HP at 2,400 rpm. Other features of the new power plant include thermostatic control of the coolant temperature, forced-feed lubrication through a gear pump at the front of the motor, a gearbox attached to the motor rather than mounted amidships, as before, and a floating rear axle with worm bevel gears replacing the former straight bevels. The engine was refined for 1923 with a new split crankshaft that introduced the (now standard) 90° offset for each pair of cylinders. Power was up to 83.5 hp. The L-Head was on the Ward's 10 Best Engines of the 20th Century list.

V10 Engine

A V10 engine is a V engine with 10 cylinders in two banks of five. The V10 configuration is not an inherently balanced design like an Inline6/V12. It can be balanced with crankshaft counterweights as a odd firing 90 degree Vee engine (BMW M5, Dodge Viper). It can be balanced with a balance shaft as an even firing 72 degree engine, or with a split crankshaft journal 90 degree Vee angle (Lamborghini Gallardo, Ford 6.8 V10). Until recently V10s had rarely been a popular configuration for road cars: a V12 is only slightly more complicated and runs more smoothly and a V8 is less complex and more economical. The V10 engine was used starting 1994 in the Dodge Ram.

V12 Engine

A V12 engine is a V engine with 12 cylinders. Like a straight-6, this configuration has perfect primary and secondary balance no matter which V angle is used and therefore needs no balance shafts.

Advantages of V Engine

- Smallest of all the eight-cylinder engines.
- The low position of bonnet helps in better aerodynamics.
- Increases primary balance & reduces vibrations.
- The smoother operation for high-speed performance.
- Compactness makes it suitable for high-end sports-cars.

W Type Engine

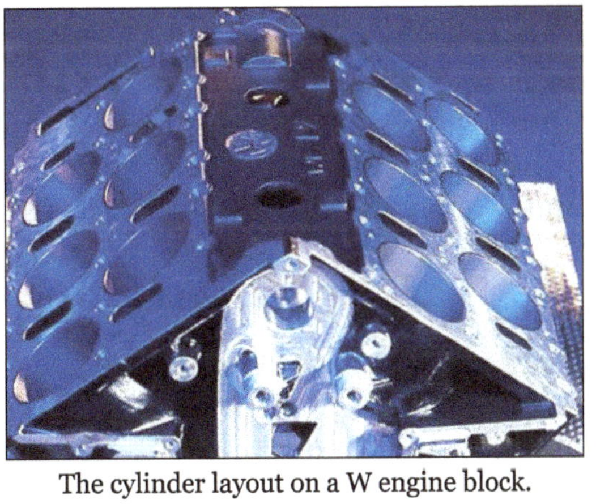

The cylinder layout on a W engine block.

The W engine is a regular piston based internal combustion four-stroke engine with the W describing the configuration of the pistons. It is another type of engine in addition to the inline and v engines. A W engine viewed from the front looks like a fat V engine and has a good reason for being so. The letter "W" is written as two V's joined together at the top but the W engine is two V's joined at the bottom so the letter and the engine although slightly different still shares a similar concept. In essence the engine is two narrow angle V engines sharing the same crankshaft. A regular V engine is between 60 to 90 degrees but the V's in the W engine is about 15 degrees so you don't see the separation between each block of the narrow angle V. The purpose of the W engine is to make a large engine further compact and the popular W engines are very large engines.

The 6 litre W-12 found in the Audi A8 and the 8 litre W-16 found in the Bugatti Veyron are two very large engines and they occupy less space by using the W configuration. The 6 litre W-12 engine in the Audi A8 will be shorter than a 6 litre V-12 found in its competitors such as the BMW 7 Series or the Mercedes S-Class. The W-12 engine will be slightly wider than the V-12 engines but it is still narrow enough to fit comfortably in the engine bay. The narrow angle allows each V to form one block with one engine head each. There may even be one set of camshafts for each V so you can understand how the W engine can resemble a V engine. The cylinders are also not directly opposite to each other but they're in a noticeable zigzag pattern. The W engine was first developed by Volkswagen for their racing cars and Volkswagen later decided to use them in production vehicles. The W configuration have been proven to be useful as there is more space to have all wheel drive parts under the bonnet as opposed to an engine of similar proportions in the V configuration.

W3 Engines

1906 Anzani 3-cylinder motorcycle engine.

One of the first W engines was the Anzani 3-cylinder, built in 1906, to be used in Anzani motorcycles. It is this W3 engine which also powered the 1909 Blériot XI, the first airplane to fly across the English Channel. The Feuling W3 is a 2.5 L (153 cu in) motorcycle engine that was built by an aftermarket parts company in the United States in the early 2000s. Like radial aircraft engines it has a master connecting rod and two slave rods connected to the pistons.

W6 Engines

The Rumpler Tropfenwagen had a Siemens and Halske-built 2,580 cc (157 cu in) overhead valve W6 engine, with three banks of paired cylinders, all working on a common crankshaft.

W8 Engines

The sole W8 engine to reach production is the Volkswagen Group W8 engine automotive engine, which used a four-bank design and was produced from 2001-2004.

W12 Engines

W12 engines with three banks of four cylinders were used by several aircraft engines from 1917 until the 1930s. A three-bank design was also used for an unsuccessful W12 engine which was intended to compete in Formula One in 1990. Although less commonly used in automobiles than V12 engines, a W12 petrol engine has been produced by Volkswagen Group since 2001. This four-bank engine— based on two VR6 engines with a common crankshaft— has been used in various cars sold under the Audi, Bentley and Volkswagen brands.

Napier Lion VII.

W16 Engines

W16 engines are rarely produced, with the notable exception of the Volkswagen Group 8.0 WR16 engine that has been used since 2005. This type of engine can also be found in the Bugatti Veyron, Chiron, Divo and their related models.

Higher Configurations

W18 Engines

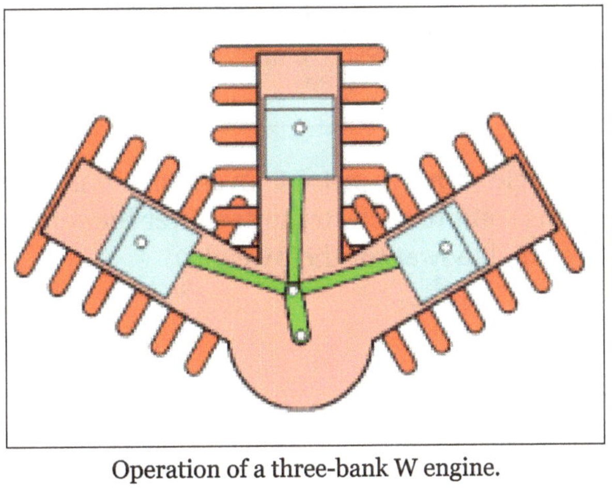

Operation of a three-bank W engine.

The W18 layout is rarely used, with the only production examples being several aircraft during the 1920s and 1930s, and CRM Motori SpA marine engines.

W24 Engine

The Allison V-3420, manufactured by the Allison Engine Company, is an example of a W24 engine.

W30 Engine

The Chrysler A57 multibank, essentially five 4.1 liter Chrysler flathead engines on one crankshaft, saw action during World War II. Developed in 1941, it would be used inside variants of the M3 Lee and M4 Sherman tanks, all deployed on the Western Front.

Opposed Piston Engine

OPEs is also known as OPOS i.e. Opposed cylinder opposed piston. OPEs combine two pistons per cylinder, working in opposite, reciprocating motion. This eliminates the cylinder head and valvetrain—considered among the most complex and costly components in conventional engines and the primary contributors to heat and friction losses. Instead of employing poppet valves for the gas exchange, OPEs are piston-ported with intake and exhaust ports located at either end of the cylinder enabling efficient uniflow air scavenging. The porting arrangement of the OPE architecture allows for the two-stroke combustion cycle, which produces twice as many power strokes per crankshaft revolution as the four-stroke engine—resulting in smaller displacement engines for similar performance and lower in-cylinder pressure to reduce emissions. Combined with a direct injection fuel system, OPEs represent an ideal platform for a compression ignition engine.

Working of OPOC Engine

The engine makes a complete cycle with two piston movements. It intakes and exhausts simultaneously during the initiation of the compression stroke and after the combustion stroke. This process allows the engines to give very high power for their weight.

Specifications of OPOC

- Dry weight: 296 pounds.
- Dimensions: 22.8 x 41.3 x 18.5 (inches), Length x Width x Height.
- Cylinder bore: 100 millimeters.
- 325 horsepower at 3500 rpm.
- Torque: 664-pound-feet at 2100 rpm.
- Power density: 1.1 horsepower/pound.

Features of OPOC

- Smaller size
- Lighter weight
- Fewer parts
- Higher efficiency
- Unmatched power density
- A truly modular design

External Combustion Engine

An external combustion engine is a heat engine where an (internal) working fluid is heated by combustion of an external source, through the engine wall or a heat exchanger. The fluid then, by expanding and acting on the mechanism of the engine produces motion and usable work. The fluid is then cooled, compressed and reused in a closed cycle. Unlike the steam engine's use of water in both its liquid and gaseous phases as the working fluid, the Stirling engine encloses a fixed quantity of permanently gaseous fluid such as air or helium. As in all heat engines, the general cycle consists of compressing cool gas, heating the gas, expanding the hot gas, and finally cooling the gas before repeating the cycle. There are two main families of external combustion engines; steam engines which rely on expanding steam (or occasionally some other vapour) to drive a mechanism; or Stirling engines which use hot air (or some other hot gas). The use of both technologies reached their zeniths around 1900 and have declined almost to extinction since. However a brief description is worthwhile, since:

- They were successfully and widely used in the past for pumping water.
- They both have the merit of being well suited to the use of low cost fuels such as coal, peat and biomass.
- Attempts to update and revive them are taking place.

And therefore they may re-appear as viable options in the longer term future. The primary disadvantage of e.c. engines is that a large area of heat exchanger is necessary to transmit heat into the working cylinder(s) and also to reject heat at the end of the cycle. As a result, e.c. engines are generally bulky and expensive to construct compared with i.c. engines. Also, since they are no longer generally manufactured they do not enjoy the economies of mass-production available to i.e. engines. They also will not start so quickly or conveniently as an i.c. engine; because it takes time to light the fire and heat the machine to its working temperature. Due to their relatively poor power/weight ratio and also the worse energy/weight ratio of solid fuels, the kinds of applications where steam or Stirling engines are most likely to be acceptable are for static applications such as as irrigation water pumping in areas where petroleum fuels are not readily available but low cost solid fuels are. On the positive side, e.c. engines have the advantage of having the potential to be much longer-lasting than i.c. engines (100 year old steam railway locomotives are relatively easy to keep in working order, but it is rare for i.c. engines to be used more than 20 years or so. E.c. engines are also significantly quieter and free of vibrations than i.c. engines. The level of skill needed for maintenance may also be lower, although the amount of time spent will be higher, particularly due to the need for cleaning out the furnace.

Modern engineering techniques promise that any future steam or Stirling engines could benefit from features not available over 60 years ago when they were last in general use. Products incorporating these new developments are not yet on the market, but R&D is in hand in various countries on a limited scale; however it will probably be some years before a new generation of multi-fuel Stirling or steam powered pumps become generally available.

References

- Rotary-engine, technology: britannica.com, Retrieved 05, April 2020
- What-is-a-diesel-engine-How-does-it-work-360001871417: support.wellandpower.net, Retrieved 12, June 2020
- Diesel-cycle-diesel-engine, thermodynamic-cycles, nuclear-engineering: nuclear-power.net, Retrieved 30, March 2020
- Concept-of-Six-Stroke-Engine: ijser.org, Retrieved 16, May 2020
- Two-stroke-motorcycle-engines-explained-tech-talk-by-kevin-cameron: cycleworld.com, Retrieved 28, February 2020
- What-is-a-single-cylinder-engine: wise-geek.com, Retrieved 19, August 2020
- Multi-cylinder-engines-automobile: what-when-how.com, Retrieved 06, April 2020
- Advantages-radial-engines: blog.covingtonaircraft.com, Retrieved 15, July 2020
- How-do-radial-engines-work: mechstuff.com, Retrieved 21, May 2020
- Innovative-opoc-engine-opposed-piston-engine-4280: techexplorist.com, Retrieved 27, January 2020

3

Boost Controller and Engine Sensors

A device that is used to regulate the boost level produced by a supercharged engine by regulating the air pressure carried out to the pneumatic and mechanical wastegate actuator is known as boost controller. The topics elaborated in this chapter will help in gaining a better perspective about boost controller as well as different types of engine sensors such as engine coolant temperature sensor, engine speed sensor (ESS) and oxygen sensor.

Boost Controller

A boost controller is a device used to control the amount of boost pressure that a turbocharger will provide. Most of the time a wastegate spring is chosen to provide the correct amount of boost pressure desired, and no other controller device is needed. To run more pressure most use a device called a Boost Controller, to control the amount of air that the wastegate diaphragm "sees". This in turn causes the wastegate to delay its opening and therefore causes the turbo to spin faster and increase boost. There are three main types of boost controllers that are in use today. Each of them have there pros and cons and will be discussed below:

Bleed Type Boost Controller

Like all boost controllers, a bleed type controller is installed on the wastegate vacuum line. It can be installed anywhere on that line, but usually it is best to install the controller as close to the wastegate as possible and to try and keep that vacuum line as short as possible. The way this controller works is by bleeding off pressure out of the line. The amount of air bled off is what controls the boost. Understanding a bleed type controller is best understood with example. Let's say we have a 10 PSI wastegate set-up. Under normal conditions the vehicle runs 10 PSI. If we install a bleed type boost controller on the vacuum line heading to the wastegate that bleeds off 2 PSI of air, then the Turbocharger will now makes 12 PSI. This is because the wastegate will not open till 10 PSI is reached, and the boost controller is bleeding off 2 psi, so it will take 12 PSI of pressure from the turbo to make 10 PSI to open the wastegate.

- Pros:
  - Very solid boost control. No wandering of boost pressure.
  - Controller least susceptible to boost spiking.
- Cons:
  - While it may not hinder boost response, it doesn't help spool response like the other controllers can.

Ball and Spring Type Boost Controller

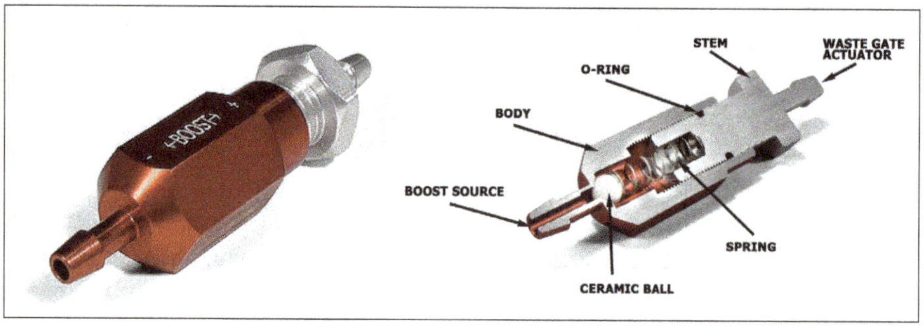

A ball and spring type boost controller is exactly what is says it is. It is a ball with a spring behind the ball. The ball blocks the flow of air until the air pressure over comes the spring. It is installed in-line on the vacuum line leading to the wastegate. Most manual Ball and spring type controllers are adjusted by tightening down the spring making it harder to open and therefore requiring a higher pressure out of the turbo to open the wastegate.

- Pros:
  - Can easily and quickly adjust boost pressures.
  - Can keep the wastegate fully closed longer during spool up, which allows the turbo to spool faster and be more responsive.
- Cons:
  - Is susceptible to over-spiking based on the design of the boost controller.

Electronic Solenoid Type Controller

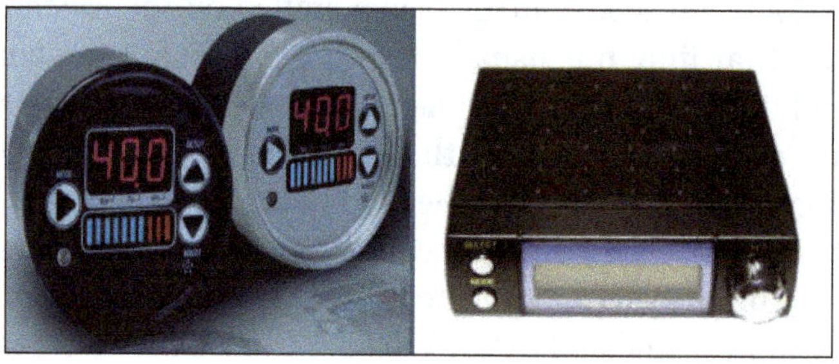

Each company has many different ways that they control the solenoid so we will just generalize how the solenoid works without getting into particulars. An electronic solenoid is installed on the wastegate vacuum line and functions like a Ball and spring type controller by hiding the boost pressure from the wastegate until a certain boost level is reached. The solenoid, being electronically controlled, is held shut until it is fed voltage by its controller and opens. The effectiveness of the solenoid is based on how well it is controlled by its computer and the software behind it.

- Pros:
  - Can be controlled from inside the vehicle.
  - Can be adjusted on the fly.

- Can increase the boost response and spool of the turbo.
- Can very the controller infinitely to help suppress boost spiking.

- Cons:
  - Set-up and programming can be daunting for beginners.

Engine Sensors

Engine sensors in a vehicle are incorporated to provide the correct amount of fuel for all operating conditions. A large number of input sensors are monitored by the engine control unit. Today, sensor technology has become common in modern vehicles. Sensors enhance safety of the people - both on board and on road, control vehicle emissions and make vehicles more efficient.

Mass Air Flow Sensor (MAF)

The MAF sensor (electric sensor) is an integral part of the engine system. It is controlled by a computer. It is located in a plastic covering between the engine and the air filter. The purpose of MAF is to calculate the amount of air intake by the engine, in terms of volume and density. For measuring the volume and density of air, the sensor uses either a hot wire or a heated filament. After the measurement, it sends a voltage signal to the computer. With this, the computer can calculate the right amount of fuel needed to maintain the correct fuel mixture for every operating condition. If there is any fault in the MAF sensor, it may result in rough idle, stalling and poor fuel economy.

Throttle Position Sensor (TPS)

The Throttle Position Sensor (TPS) is a variable resistor attached or mounted on the throttle body and is operated by moving along with the throttle shaft or spindle. The TPS changes the resistances as the throttle opens and closes, and sends a voltage signal to the computer showing the angle or position of the throttle. Thus, the TPS causes the Electronic Control Unit (ECU) to use the data to measure the engine load, fuel delivery adjust timing, acceleration, deceleration when the engine is idle or in wide open throttle, and then makes the changes according to the operating conditions. Fuel rate is either increased or decreased to achieve this.

Coolant Temperature Sensor (CTS)

The Coolant Temperature Sensor (CTS) is a temperature dependent variable resistor located on the cylinder head or intake manifold. The CTS is an important sensor and the operating strategy of the engine depends on the signal it sends. So, it is called the "master" sensor. The CTS measures the internal temperature of the engine coolant. It also senses the changes in temperature and sends a voltage signal to the Power train Control Module (PCM) for determining whether the engine is cold or warming up, is at normal operating temperature or is overheating.

Oxygen Sensor

The oxygen sensor is located on the exhaust manifold. This sensor monitors the amount of unburned oxygen present in the exhaust. When the fuel mixture is rich, most of the oxygen is exhausted during the combustion. So, only a little unburned oxygen will be left out in the exhaust. Difference in the oxygen levels creates an electrical potential, which causes the sensor to generate a voltage signal. This helps the ECU to check the quality of fuel mixture to make the changes accordingly. The sensor output will be high if the fuel mixture is rich, and the sensor output will be low if the fuel mixture is lean.

Manifold Absolute Pressure Sensor (MAP)

The MAP is a key sensor as it senses the engine load. It is mounted on the intake manifold. It monitors the difference between the air pressure in the intake manifold and outside. This sensor responds to the vacuum in the intake manifold and generates a voltage signal accordingly. It then sends the signal to the PCM. The input of the sensor is used for adjusting the fuel mixture and ignition timing, according to the changes.

Engine Speed Sensor (ESS)

The ESS is a sensor attached to the crankshaft of the car's engine. It is different from vehicle speed sensor. The ESS is used for monitoring the engine speed. In other words, it is meant for assessing the speed at which the crankshaft spins.

Voltage Sensor

The voltage sensor monitors the system voltage of the vehicle and reports it to PCM so that it can rise the idle speed of the vehicle, if the voltage is dropping. Engine sensors are an important technological innovation. They lead to better performance, better quality and more years of driving experience.

Engine Coolant Temperature Sensor

The engine coolant temperature sensor is temperature-variable resistor, which usually has a negative temperature coefficient. It is a two-wire thermistor immersed in coolant and measures its temperature. The onboard computer uses the signal of ECT as the main correction factor when calculating the ignition advance and the injection duration.

Appearance

The ECT sensor is shown on figure.

Working Principle of the ECT Sensor

In order to convert the ECT resistance variation to voltage variation, which is further processed by the ECU, the ECT sensor is connected in a circuit typically supplied with a reference voltage of +5V. In cold engine and an ambient temperature of 20 °C the sensor resistance is between 2000Ω and 3000Ω. After the engine start, coolant temperature begins to rise. ECT gradually heats and its resistance reduces proportionately. At 90 °C its resistance is in the range of 200Ω to 300Ω. Thereby, a coolant temperature dependent variable voltage signal is send to the onboard computer.

Types of ECT Sensors

- With negative temperature coefficient. These are the most popular sensors used in cars. Their resistance decreases with the increase of the temperature.
- With positive temperature coefficient. Used in some older systems, such as Renix. Here the voltage and the resistance are increasing with the raise of the temperature.

Procedure for Functionality Verification of the ECT Sensor

Sensor with negative temperature coefficient - Test with voltmeter:

- Open the protective rubber cap on the connector of the cooling system temperature sensor.
- Connect the negative wire of the voltmeter to the chassis ground.
- Determine which are the signal and the ground terminals.
- Connect the positive wire of the voltmeter is to the ECT signal terminal.
- Start the engine form cold state.
- Depending on the temperature, voltage readings should be in the range of 2V to 3V. The relationship between the voltage and the temperature is shown in table.
- Check whether the ECT voltage signal corresponds to the temperature. You will need a thermometer for the purpose.
- Start the engine and warm it to the operating temperature. During engine warming the voltage should decrease in accordance with the values given in table.
- Common problem is that the output resistance (and voltage) changes incorrectly beyond its normal range. The normal value of the ECT sensor voltage is 2V at cold engine and 0.5V at warm engine. A defect sensor can indicate a voltage of 1.5V at cold engine and 1.25V at warm engine, thus causing difficulties of starting a cold engine and a presence of rich fuel mixture when the engine is warm. This will not generate any fault codes (if the onboard controller is not programmed to detect voltage changes) since the sensor continues to operate within its design parameters. If such defect is found, the ECT sensor must be replaced.
- If the ECT voltage signal is equal to 0V (lack of power supply or there is a short circuit to the ground) or if it is 5.0V - we have an open circuit.

Temperature, °C	Resistance, Ω	Voltage, V
0	4800 - 6600	4.00 – 4.50
10	4000	3.75 – 4.00
20	2200 - 2800	3.00 – 3.50
30	1300	3.25
40	1000 – 1200	2.50 – 3.00
50	1000	2.50
60	800	2.00 – 2.50
80	270 - 380	1.00 – 1.30
110		0.50
Broken circuit		5.0 ± 0.1
Short circuit to the ground		0

Possible Sensor Failures

- The voltage of the ECT signal terminal is equal to 0V.
    - Check sensor terminals for short circuit to ground.
    - Check the integrity of the signal wires between the sensor and the onboard controller.
    - If all wires are correct, but there is no output voltage from the onboard controller, you have to verify all power supply and ground connections of the onboard controller. If supply voltages and grounds are good, the onboard controller itself falls under suspicion.
- The voltage of ECT signal terminal is equal to 5.0V. The voltage has such a value when an open circuit is present and may be obtained in one of the following conditions:
    - The signal terminal of the ECT sensor does not provide connection to the sensor.
    - Sensor circuit is open.
    - Sensor's ground circuit is open.
- The voltage signal or the reference voltage is equal to the car battery voltage.
    - Check short circuit in the wire, connected to the positive terminal of the car battery or the power supply wire.

Sensor with negative temperature coefficient - Check with ohmmeter with ECT sensor detached from the car:

- Place the sensor in a suitable water container and measure the water temperature.
- Measure the sensor's resistance and compare it with the values given in table, showing the relationship between the resistance and the temperature.
- Heat the water and take measurements periodically as well as the resistance of the sensor. Compare the results with those in table.

Sensor with negative temperature coefficient - Voltage measurement with oscilloscope:

- Connect the active end of the oscilloscope probe to the signal terminal of the sensor and the ground probe - to chassis ground.
- Set the trigger sweep of the oscilloscope in a continuous measurement mode (registration of slowly changing signals).
- Place the sensor in a suitable heated water container. After a few minutes of measurement, during the water heating time, oscilloscope screen will show the sensor's voltage change curve. Pay attention to the measuring time - it is about 10 minutes.
- It is desirable to continuously measure the temperature of the heated water with a thermometer and compare it with the values listed in table.

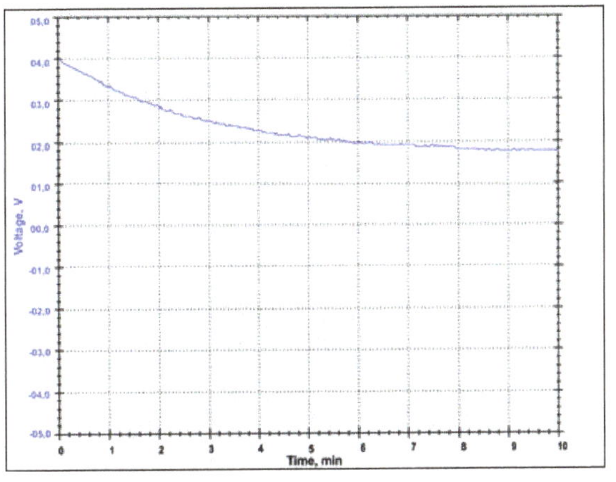

Sensor with a Positive Temperature Coefficient

ECT sensor with a positive temperature coefficient of resistance is a thermistor whose resistance increases with the rising of the temperature. It is used in a small number of systems (mainly in Renault cars). The overall method of verification is similar to the method for checking a sensor with negative temperature coefficient, described above. Received measurement data can be compared with the data given in table showing the relationship between the resistance and the temperature of the sensor.

Table: Relationship between the resistance and the temperature of the sensor.

Temperature, °C	Resistance, Ω	Voltage, V
0	254 - 266	
20	283 - 297	0.6 – 0.8
80	383 - 397	1.0 – 1.2
Open circuit		5.0 ± 0.1
Short circuit to ground		0

Engine Speed Sensor (ESS)

The engine speed sensor is the most important sensor of the engine management system. In addition to the speed, this sensor, together with the trigger wheel determines the crankshaft position.

In addition to a crankshaft position sensor, a camshaft phase sensor can also be used. The sensors are available in 3 variants.

- The Variable Reluctance Sensor
- The Hall Effect Sensor
- The OPTO Sensor

The VR Sensor

This sensor consists of a magnet around which a coil is wound. By moving a piece of metal towards the sensor, the magnetic field will change. The same happens when we pull the metal away from the sensor. The changing magnetic field in the coil of the sensor will generate a voltage. If the metal object moves towards it, the voltage will be positive, if the metal object moves away from it, the voltage will be negative. The signal coming from the sensor is therefore a varying positive and negative voltage. An alternating voltage. We see a new sinus for every tooth of the trigger wheel. The voltage generated by this sensor differs. At starting speed this will be approximately 1 Volt (measured in the AC position). This can be as high as 100 volts if the engine makes a lot of revs.

VR sensor

The HALL Effect Sensor

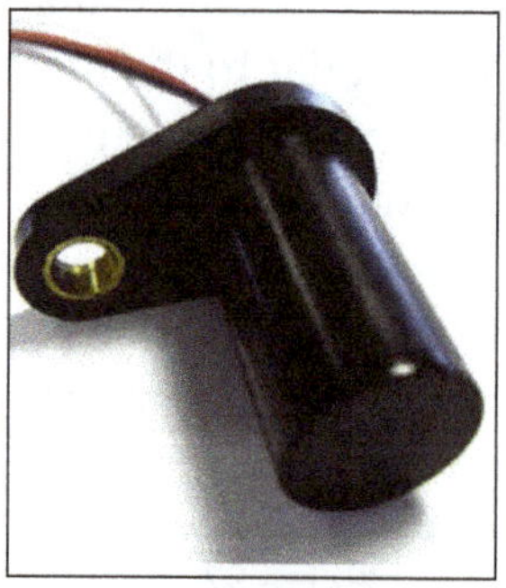

HALL sensor

Responds to magnetism. This sensor has its own magnet, but also a piece of electronics that responds to the proximity of a magnet. In the case of a HALL sensor with built-in magnet, the metal of the trigger wheel ensures that the magnetism reaches the sensor. Most HALL sensors will switch to ground if there is metal nearby. This signal is interrupted if there is no metal nearby. The sensor therefore produces no sine wave and the voltage cannot be measured. A "pull-up" resistor is required to make a switching signal.

VR or HALL Sensor

We usually use a VR sensor as a crankshaft sensor. As a camshaft phase sensor we usually use a HALL sensor. We can sometimes see the difference between those sensors, but you can defenetly measure it. A HALL sensor always has 3 connections. Namely a power supply (+), a mass (-) and a signal (0). A VR sensor sometimes has 2 connections and if it is a type with a wire it usually has 3 connections. You can measure the coil between 2 connections. This will give a resistance of between 150 and 1200 ohms. You do not measure anything on the third thread. This is the shielding of the wire. The shield ensures that no interference can occur in the signal due to influences from other wiring. With the ECU, this shield must be grounded. We measure much higher resistance values with a HALL sensor.

OPTO Sensor

This is a light lock sensor. This is found in some Japanese cars. For example in the Mitsubishi 4G63 system, used in the first Mazda MX-5, among others. In terms of connection in identification, it behaves the same as the HALL sensor.

OPTO sensor

Oxygen Sensor

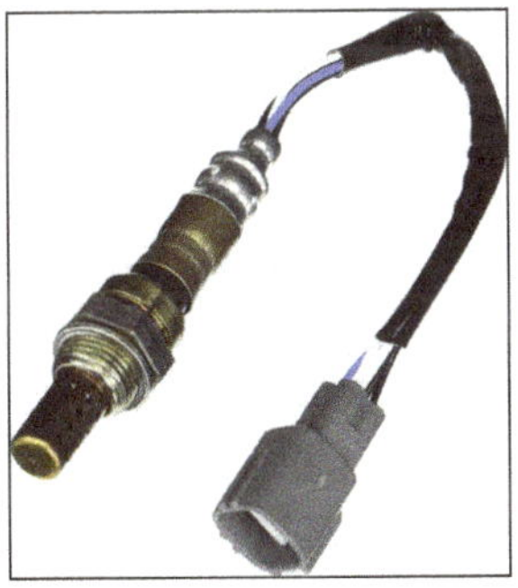

Oxygen sensor

An oxygen sensor is one type of sensor and it is available in the exhaust system of an automobile. The size and shape of this sensor look like a spark plug. Based on its arrangement in regard to the catalytic converter, this sensor can be arranged before (upstream) or after (downstream) the converter. Most of the automobiles which are designed after 1990 include upstream & downstream 02 sensors. The oxygen sensors used in automobiles are one sensor is arranged in front of the catalytic converter & one is arranged in every exhaust manifold of the automobile. But, the maximum number of these sensors in a car mainly depends on the engine, model and year. But, most of the vehicles have 4-sensors.

- Working Principle: The working principle of the o2 sensor is to check the oxygen amount within the exhaust. Firstly, this oxygen was added to the fuel for good ignition. The communication of this sensor can be done with the help of a voltage signal. So the oxygen status in the exhaust will be decided by the computer of the car. The computer regulates the mixture of fuel or oxygen delivered to the car engine. The arrangement of the sensor before & after the catalytic converter permits to maintain the hygiene of the exhaust & check the converter's efficiency.
- Types of Oxygen Sensors: Oxygen sensors are classified into two namely binary exhaust gas and universal exhaust gas.
    - Binary Exhaust Gas Oxygen Sensor: The binary sensor gives a transition within electric voltage at 350 °C temperature based on the level of oxygen within the exhaust. It contrasts the remaining oxygen content within the exhaust by the ambient air oxygen level & recognizes the change from a lack of air to an excess air & vice versa.
    - Universal Exhaust Gas: This sensor is very exact when calculating the ratios of lack and excess of air or fuel. It has a better calculating range & is also appropriate for employ in gas & diesel engines.

Signs of Faulty Sensor

The faulty sensor can be found by using the following signs:

- Breakdown to exceed the emissions analysis.
- Fuel mileage can be decreased.
- The engine light will be off.
- Performance is poor, stalling and rough idling.
- Code checker recognizing sensor failure.

Applications

The applications of oxygen sensors include Marine Respiration, Quick Reaction Monitoring, Benchtop R&D, Monitoring of Fuel Tank, Particular Hydrocarbon Environments, Long-term Procedure Monitoring, Fermentation, Packing of Food & Beverage Packaging, Pharmaceutical & Medical, etc.

References

- Boost-Controller-Types: redline-motorsports.net, Retrieved 13, March 2020
- Types-of-engine-sensors, mechanical-engineering: idc-online.com, Retrieved 22, June 2020
- Engine-coolant-temperature-sensor-ect-13-1: autoditex.com, Retrieved 26, August 2020
- Engine-speed-sensor-3632494: efignition.com, Retrieved 11, May 2020
- Oxygen-sensor-working-and-applications: elprocus.com, Retrieved 02, July 2020

4

Mechanics of Combustion

Combustion refers to the exothermic redox chemical reaction that takes place between a fuel and an oxidant giving off heat and light in the process. Some concepts related to combustion such as combustion triangle, fuel, thermodynamics of combustion, stoichiometry and conservation of mass are discussed in this chapter in an easy-to-understand manner.

Combustion

The flames caused as a result of a fuel undergoing combustion (burning).

Combustion or burning is a high-temperature exothermic redox chemical reaction between a fuel (the reductant) and an oxidant, usually atmospheric oxygen, that produces oxidized, often gaseous products, in a mixture termed as smoke. Combustion in a fire produces a flame, and the heat produced can make combustion self-sustaining. Combustion is often a complicated sequence of elementary radical reactions. Solid fuels, such as wood, first undergo endothermic pyrolysis to produce gaseous fuels whose combustion then supplies the heat required to produce more of them. Combustion is often hot enough that light in the form of either glowing or a flame is produced. A simple example can be seen in the combustion of hydrogen and oxygen into water vapor, a reaction commonly used to fuel rocket engines. This reaction releases 242 kJ/mol of heat and reduces the enthalpy accordingly (at constant temperature and pressure):

Air pollution abatement equipment provides combustion control for industrial processes.

$$2H_2(g) + O_2(g) \rightarrow 2H_2O(g)$$

Combustion of an organic fuel in air is always exothermic because the double bond in O_2 is much weaker than other double bonds or pairs of single bonds, and therefore the formation of the stronger bonds in the combustion products CO_2 and H_2O results in the release of energy. The bond energies in the fuel play only a minor role, since they are similar to those in the combustion products; e.g., the sum of the bond energies of CH_4 is nearly the same as that of CO_2. The heat of combustion is approximately -418 kJ per mole of O_2 used up in the combustion reaction, and can be estimated from the elemental composition of the fuel.

Uncatalyzed combustion in air requires fairly high temperatures. Complete combustion is stoichiometric with respect to the fuel, where there is no remaining fuel, and ideally, no remaining oxidant. Thermodynamically, the chemical equilibrium of combustion in air is overwhelmingly on the side of the products. However, complete combustion is almost impossible to achieve, since the chemical equilibrium is not necessarily reached, or may contain unburnt products such as carbon monoxide, hydrogen and even carbon (soot or ash). Thus, the produced smoke is usually toxic and contains unburned or partially oxidized products. Any combustion at high temperatures in atmospheric air, which is 78 percent nitrogen, will also create small amounts of several nitrogen oxides, commonly referred to as NO_x, since the combustion of nitrogen is thermodynamically favored at high, but not low temperatures. Since combustion is rarely clean, flue gas cleaning or catalytic converters may be required by law.

Fires occur naturally, ignited by lightning strikes or by volcanic products. Combustion (fire) was the first controlled chemical reaction discovered by humans, in the form of campfires and bonfires, and continues to be the main method to produce energy for humanity. Usually, the fuel is carbon, hydrocarbons or more complicated mixtures such as wood that contains partially oxidized hydrocarbons. The thermal energy produced from combustion of either fossil fuels such as coal or oil, or from renewable fuels such as firewood, is harvested for diverse uses such as cooking, production of electricity or industrial or domestic heating. Combustion is also currently the only reaction used to power rockets. Combustion is also used to destroy (incinerate) waste, both nonhazardous and hazardous.

Oxidants for combustion have high oxidation potential and include atmospheric or pure oxygen, chlorine, fluorine, chlorine trifluoride, nitrous oxide and nitric acid. For instance, hydrogen burns in chlorine to form hydrogen chloride with the liberation of heat and light characteristic of combustion. Although usually not catalyzed, combustion can be catalyzed by platinum or vanadium, as in the contact process.

Types

Complete

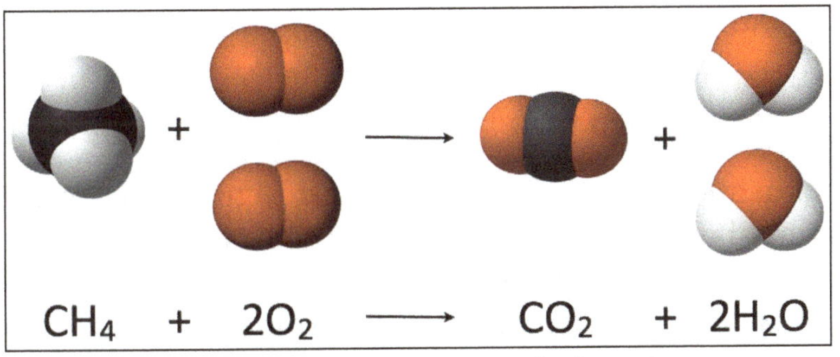

The combustion of methane, a hydrocarbon.

In complete combustion, the reactant burns in oxygen, producing a limited number of products. When a hydrocarbon burns in oxygen, the reaction will primarily yield carbon dioxide and water. When elements are burned, the products are primarily the most common oxides. Carbon will yield carbon dioxide, sulfur will yield sulfur dioxide, and iron will yield iron(III) oxide. Nitrogen is not considered to be a combustible substance when oxygen is the oxidant, but small amounts of various nitrogen oxides (commonly designated NO_x species) form when air is the oxidant.

Combustion is not necessarily favorable to the maximum degree of oxidation, and it can be temperature-dependent. For example, sulfur trioxide is not produced quantitatively by the combustion of sulfur. NOx species appear in significant amounts above about 2,800°F (1,540°C), and more is produced at higher temperatures. The amount of NOx is also a function of oxygen excess.

In most industrial applications and in fires, air is the source of oxygen (O_2). In air, each mole of oxygen is mixed with approximately 3.71 mol of nitrogen. Nitrogen does not take part in combustion, but at high temperatures some nitrogen will be converted to NO_x (mostly NO, with much smaller amounts of NO_2). On the other hand, when there is insufficient oxygen to completely combust the fuel, some fuel carbon is converted to carbon monoxide and some of the hydrogen remains unreacted. A more complete set of equations for the combustion of a hydrocarbon in air therefore requires an additional calculation for the distribution of oxygen between the carbon and hydrogen in the fuel.

The amount of air required for complete combustion to take place is known as theoretical air. However, in practice the air used is 2-3x that of theoretical air.

Incomplete

Incomplete combustion will occur when there is not enough oxygen to allow the fuel to react completely to produce carbon dioxide and water. It also happens when the combustion is quenched by a heat sink, such as a solid surface or flame trap.

For most fuels, such as diesel oil, coal or wood, pyrolysis occurs before combustion. In incomplete combustion, products of pyrolysis remain unburnt and contaminate the smoke with noxious particulate matter and gases. Partially oxidized compounds are also a concern; partial oxidation of ethanol can produce harmful acetaldehyde, and carbon can produce toxic carbon monoxide.

The quality of combustion can be improved by the designs of combustion devices, such as burners and internal combustion engines. Further improvements are achievable by catalytic after-burning devices (such as catalytic converters) or by the simple partial return of the exhaust gases into the combustion process. Such devices are required by environmental legislation for cars in most countries, and may be necessary to enable large combustion devices, such as thermal power stations, to reach legal emission standards.

The degree of combustion can be measured and analyzed with test equipment. HVAC contractors, firemen and engineers use combustion analyzers to test the efficiency of a burner during the combustion process. In addition, the efficiency of an internal combustion engine can be measured in this way, and some U.S. states and local municipalities use combustion analysis to define and rate the efficiency of vehicles on the road today.

Smouldering

Smouldering is the slow, low-temperature, flameless form of combustion, sustained by the heat evolved when oxygen directly attacks the surface of a condensed-phase fuel. It is a typically incomplete combustion reaction. Solid materials that can sustain a smouldering reaction include coal, cellulose, wood, cotton, tobacco, peat, duff, humus, synthetic foams, charring polymers (including polyurethane foam) and dust. Common examples of smouldering phenomena are the initiation of residential fires on upholstered furniture by weak heat sources (e.g., a cigarette, a short-circuited wire) and the persistent combustion of biomass behind the flaming fronts of wildfires.

Rapid

Rapid combustion is a form of combustion, otherwise known as a fire, in which large amounts of heat and light energy are released, which often results in a flame. This is used in a form of machinery such as internal combustion engines and in thermobaric weapons. Such a combustion is frequently called an explosion, though for an internal combustion engine this is inaccurate. An internal combustion engine nominally operates on a controlled rapid burn. When the fuel-air mixture in an internal combustion engine explodes, that is known as detonation.

Spontaneous

Spontaneous combustion is a type of combustion which occurs by self heating (increase in temperature due to exothermic internal reactions), followed by thermal runaway (self heating which rapidly accelerates to high temperatures) and finally, ignition. For example, phosphorus self-ignites at room temperature without the application of heat.

Turbulent

Combustion resulting in a turbulent flame is the most used for industrial application (e.g. gas turbines, gasoline engines, etc.) because the turbulence helps the mixing process between the fuel and oxidizer.

Micro-gravity

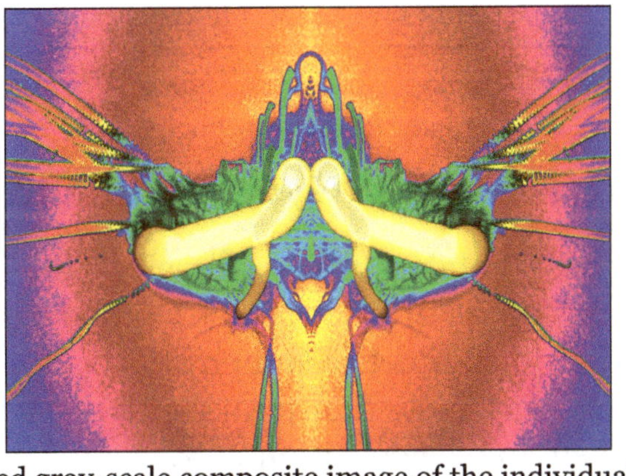

Colourized gray-scale composite image of the individual frames from a video of a backlit fuel droplet burning in microgravity.

The term 'micro' gravity refers to a gravitational state that is 'low' (i.e., 'micro' in the sense of 'small' and not necessarily a millionth of Earth's normal gravity) such that the influence of buoyancy on physical processes may be considered small relative to other flow processes that would be present at normal gravity. In such an environment, the thermal and flow transport dynamics can behave quite differently than in normal gravity conditions (e.g., a candle's flame takes the shape of a sphere.). Microgravity combustion research contributes to the understanding of a wide variety of aspects that are relevant to both the environment of a spacecraft (e.g., fire dynamics relevant to crew safety on the International Space Station) and terrestrial (Earth-based) conditions (e.g., droplet combustion dynamics to assist developing new fuel blends for improved combustion, materials fabrication processes, thermal management of electronic systems, multiphase flow boiling dynamics, and many others).

Micro-combustion

Combustion processes which happen in very small volumes are considered micro-combustion. The high surface-to-volume ratio increases specific heat loss. Quenching distance plays a vital role in stabilizing the flame in such combustion chambers.

Chemical Equations

Stoichiometric Combustion of a Hydrocarbon in Oxygen

Generally, the chemical equation for stoichiometric combustion of a hydrocarbon in oxygen is:

$$C_xH_y + zO_2 -> xCO_2 + \frac{y}{2}H_2O$$

where $z = x + y/4$.

For example, the stoichiometric burning of propane in oxygen is:

$$\underbrace{C_3H_8}_{\substack{propane \\ (fuel)}} + \underbrace{5O_2}_{oxygen} \rightarrow \underbrace{3CO_2}_{carbon\ dioxide} + \underbrace{4H_2O}_{water}$$

Stoichiometric Combustion of a Hydrocarbon in Air

If the stoichiometric combustion takes place using air as the oxygen source, the nitrogen present in the air (Atmosphere of Earth) can be added to the equation (although it does not react) to show the stoichiometric composition of the fuel in air and the composition of the resultant flue gas. Note that treating all non-oxygen components in air as nitrogen gives a 'nitrogen' to oxygen ration of 3.77, i.e. (100% - O_2%) / O_2% where O_2% is 20.95% vol:

$$\mathrm{C_xH_y + zO_2 + 3.77zN_2 \rightarrow xCO_2 + \frac{y}{2}H_2O + 3.77zN_2}$$

where $z = x + \frac{1}{4}y$.

For example, the stoichiometric combustion of propane (C3H8) in air is:

$$\underbrace{\mathrm{C_3H_8}}_{\text{fuel}} + \underbrace{\mathrm{5O_2}}_{\text{oxygen}} + \underbrace{\mathrm{18.87N_2}}_{\text{nitrogen}} \rightarrow \underbrace{\mathrm{3CO_2}}_{\text{carbon dioxide}} + \underbrace{\mathrm{4H_2O}}_{\text{water}} + \underbrace{\mathrm{18.87N_2}}_{\text{nitrogen}}$$

The stoichiometric composition of propane in air is 1 / (1 + 5 + 18.87) = 4.02% vol

Trace Combustion Products

Various other substances begin to appear in significant amounts in combustion products when the flame temperature is above about 1600 K. When excess air is used, nitrogen may oxidize to NO and, to a much lesser extent, to NO_2. CO forms by disproportionation of CO_2, and H_2 and OH form by disproportionation of H_2O.

For example, when 1 mol of propane is burned with 28.6 mol of air (120% of the stoichiometric amount), the combustion products contain 3.3% O_2. At 1400 K, the equilibrium combustion products contain 0.03% NO and 0.002% OH. At 1800 K, the combustion products contain 0.17% NO, 0.05% OH, 0.01% CO, and 0.004% H_2.

Diesel engines are run with an excess of oxygen to combust small particles that tend to form with only a stoichiometric amount of oxygen, necessarily producing nitrogen oxide emissions. Both the United States and European Union enforce limits to vehicle nitrogen oxide emissions, which necessitate the use of special catalytic converters or treatment of the exhaust with urea.

Incomplete Combustion of a Hydrocarbon in Oxygen

The incomplete (partial) combustion of a hydrocarbon with oxygen produces a gas mixture containing mainly CO_2, CO, H_2O, and H_2. Such gas mixtures are commonly prepared for use as protective atmospheres for the heat-treatment of metals and for gas carburizing. The general reaction equation for incomplete combustion of one mole of a hydrocarbon in oxygen is:

$$\underbrace{C_xH_y}_{fuel} + \underbrace{zO_2}_{oxygen} \rightarrow \underbrace{aCO_2}_{carbon\ dioxide} + \underbrace{bCO}_{carbon\ monoxide} + \underbrace{cH_2O}_{water} + \underbrace{dH_2}_{hydrogen}$$

When *z* falls below roughly 50% of the stoichiometric value, CH4 can become an important combustion product; when *z* falls below roughly 35% of the stoichiometric value, elemental carbon may become stable.

The products of incomplete combustion can be calculated with the aid of a material balance, together with the assumption that the combustion products reach equilibrium. For example, in the combustion of one mole of propane (C_3H_8) with four moles of O_2, seven moles of combustion gas are formed, and *z* is 80% of the stoichiomctric valuc. The three elemental balance equations are:

- Carbon: $a+b=3$
- Hydrogen: $2c+2d=8$
- Oxygen: $2a+b+c=8$

These three equations are insufficient in themselves to calculate the combustion gas composition. However, at the equilibrium position, the water-gas shift reaction gives another equation:

$$CO + H_2O -> CO_2 + H_2 ; \; K_{eq} = \frac{a \times d}{b \times c}$$

For example, at 1200 K the value of K_{eq} is 0.728. Solving, the combustion gas consists of 42.4% H_2O, 29.0% CO_2, 14.7% H_2, and 13.9% CO. Carbon becomes a stable phase at 1200 K and 1 atm pressure when z is less than 30% of the stoichiometric value, at which point the combustion products contain more than 98% H_2 and CO and about 0.5% CH_4.

Fuels

Substances or materials which undergo combustion are called fuels. The most common examples are natural gas, propane, kerosene, diesel, petrol, charcoal, coal, wood, etc.

Liquid Fuels

Combustion of a liquid fuel in an oxidizing atmosphere actually happens in the gas phase. It is the vapor that burns, not the liquid. Therefore, a liquid will normally catch fire only above a certain temperature: its flash point. The flash point of a liquid fuel is the lowest temperature at which it can form an ignitable mix with air. It is the minimum temperature at which there is enough evaporated fuel in the air to start combustion.

Gaseous Fuels

Combustion of gaseous fuels may occur through one of four distinctive types of burning: diffusion flame, premixed flame, autoignitive reaction front, or as a detonation. The type of burning that actually occurs depends on the degree to which the fuel and oxidizer are mixed prior to heating: for example, a diffusion flame is formed if the fuel and oxidizer are separated initially, whereas a premixed flame is formed otherwise. Similarly, the type of burning also depends on the pressure: a detonation, for example, is an autoignitive reaction front coupled to a strong shock wave giving it its characteristic high-pressure peak and high detonation velocity.

Solid Fuels

The act of combustion consists of three relatively distinct but overlapping phases:

- Preheating phase, when the unburned fuel is heated up to its flash point and then fire point. Flammable gases start being evolved in a process similar to dry distillation.
- Distillation phase or gaseous phase, when the mix of evolved flammable gases with oxygen is ignited. Energy is produced in the form of heat and light. Flames

are often visible. Heat transfer from the combustion to the solid maintains the evolution of flammable vapours.

- Charcoal phase or solid phase, when the output of flammable gases from the material is too low for persistent presence of flame and the charred fuel does not burn rapidly and just glows and later only smoulders.

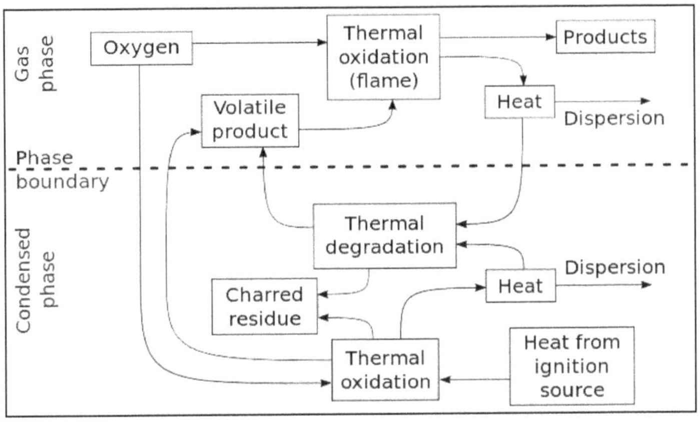

A general scheme of polymer combustion.

Combustion Management

Efficient process heating requires recovery of the largest possible part of a fuel's heat of combustion into the material being processed. There are many avenues of loss in the operation of a heating process. Typically, the dominant loss is sensible heat leaving with the offgas (i.e., the flue gas). The temperature and quantity of offgas indicates its heat content (enthalpy), so keeping its quantity low minimizes heat loss.

In a perfect furnace, the combustion air flow would be matched to the fuel flow to give each fuel molecule the exact amount of oxygen needed to cause complete combustion. However, in the real world, combustion does not proceed in a perfect manner. Unburned fuel (usually CO and H_2) discharged from the system represents a heating value loss (as well as a safety hazard). Since combustibles are undesirable in the offgas, while the presence of unreacted oxygen there presents minimal safety and environmental concerns, the first principle of combustion management is to provide more oxygen than is theoretically needed to ensure that all the fuel burns. For methane (CH_4) combustion, for example, slightly more than two molecules of oxygen are required.

The second principle of combustion management, however, is to not use too much oxygen. The correct amount of oxygen requires three types of measurement: first, active control of air and fuel flow; second, offgas oxygen measurement; and third, measurement of offgas combustibles. For each heating process there exists an optimum condition of minimal offgas heat loss with acceptable levels of combustibles concentration. Minimizing excess oxygen pays an additional benefit: for a given offgas temperature, the NOx level is lowest when excess oxygen is kept lowest.

Adherence to these two principles is furthered by making material and heat balances on the combustion process. The material balance directly relates the air/fuel ratio to the percentage of O_2 in the combustion gas. The heat balance relates the heat available for the charge to the overall net heat produced by fuel combustion. Additional material and heat balances can be made to quantify the thermal advantage from preheating the combustion air, or enriching it in oxygen.

Reaction Mechanism

Combustion in oxygen is a chain reaction in which many distinct radical intermediates participate. The high energy required for initiation is explained by the unusual structure of the dioxygen molecule. The lowest-energy configuration of the dioxygen molecule is a stable, relatively unreactive diradical in a triplet spin state. Bonding can be described with three bonding electron pairs and two antibonding electrons, whose spins are aligned, such that the molecule has nonzero total angular momentum. Most fuels, on the other hand, are in a singlet state, with paired spins and zero total angular momentum. Interaction between the two is quantum mechanically a "forbidden transition", i.e. possible with a very low probability. To initiate combustion, energy is required to force dioxygen into a spin-paired state, or singlet oxygen. This intermediate is extremely reactive. The energy is supplied as heat, and the reaction then produces additional heat, which allows it to continue.

Combustion of hydrocarbons is thought to be initiated by hydrogen atom abstraction (not proton abstraction) from the fuel to oxygen, to give a hydroperoxide radical (HOO). This reacts further to give hydroperoxides, which break up to give hydroxyl radicals. There are a great variety of these processes that produce fuel radicals and oxidizing radicals. Oxidizing species include singlet oxygen, hydroxyl, monatomic oxygen, and hydroperoxyl. Such intermediates are short-lived and cannot be isolated. However, non-radical intermediates are stable and are produced in incomplete combustion. An example is acetaldehyde produced in the combustion of ethanol. An intermediate in the combustion of carbon and hydrocarbons, carbon monoxide, is of special importance because it is a poisonous gas, but also economically useful for the production of syngas.

Solid and heavy liquid fuels also undergo a great number of pyrolysis reactions that give more easily oxidized, gaseous fuels. These reactions are endothermic and require constant energy input from the ongoing combustion reactions. A lack of oxygen or other poorly designed conditions result in these noxious and carcinogenic pyrolysis products being emitted as thick, black smoke.

The rate of combustion is the amount of a material that undergoes combustion over a period of time. It can be expressed in grams per second (g/s) or kilograms per second (kg/s).

Detailed descriptions of combustion processes, from the chemical kinetics perspective, requires the formulation of large and intricate webs of elementary reactions. For instance, combustion of hydrocarbon fuels typically involve hundreds of chemical species reacting according to thousands of reactions.

Inclusion of such mechanisms within computational flow solvers still represents a pretty challenging task mainly in two aspects. First, the number of degrees of freedom (proportional to the number of chemical species) can be dramatically large; second the source term due to reactions introduces a disparate number of time scales which makes the whole dynamical system stiff. As a result, the direct numerical simulation of turbulent reactive flows with heavy fuels soon becomes intractable even for modern supercomputers.

Therefore, a plethora of methodologies has been devised for reducing the complexity of combustion mechanisms without renouncing to high detail level. Examples are provided by: the Relaxation Redistribution Method (RRM) The Intrinsic Low-Dimensional Manifold (ILDM) approach and further developments The invariant constrained equilibrium edge preimage curve method. A few variational approaches The Computational Singular perturbation (CSP) method and further developments. The Rate Controlled Constrained Equilibrium (RCCE) and Quasi Equilibrium Manifold (QEM) approach. The G-Scheme. The Method of Invariant Grids (MIG).

Temperature

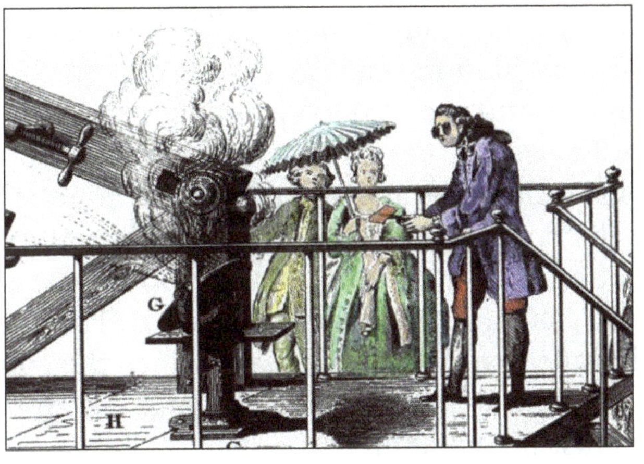

Antoine Lavoisier conducting an experiment related combustion generated by amplified sun light.

Assuming perfect combustion conditions, such as complete combustion under adiabatic conditions (i.e., no heat loss or gain), the adiabatic combustion temperature can be determined. The formula that yields this temperature is based on the first law of thermodynamics and takes note of the fact that the heat of combustion is used entirely for heating the fuel, the combustion air or oxygen, and the combustion product gases (commonly referred to as the *flue gas*).

In the case of fossil fuels burnt in air, the combustion temperature depends on all of the following:

- The heating value

- The stoichiometric air to fuel ratio λ
- The specific heat capacity of fuel and air
- The air and fuel inlet temperatures

The adiabatic combustion temperature (also known as the *adiabatic flame temperature*) increases for higher heating values and inlet air and fuel temperatures and for stoichiometric air ratios approaching one.

Most commonly, the adiabatic combustion temperatures for coals are around 2,200 °C (3,992 °F) (for inlet air and fuel at ambient temperatures and for $\lambda = 1.0$), around 2,150 °C (3,902 °F) for oil and 2,000 °C (3,632 °F) for natural gas.

In industrial fired heaters, power stationsteam generators, and large gas-fired turbines, the more common way of expressing the usage of more than the stoichiometric combustion air is *percent excess combustion air*. For example, excess combustion air of 15 percent means that 15 percent more than the required stoichiometric air is being used.

Instabilities

Combustion instabilities are typically violent pressure oscillations in a combustion chamber. These pressure oscillations can be as high as 180 dB, and long term exposure to these cyclic pressure and thermal loads reduces the life of engine components. In rockets, such as the F1 used in the Saturn V program, instabilities led to massive damage of the combustion chamber and surrounding components. This problem was solved by re-designing the fuel injector. In liquid jet engines the droplet size and distribution can be used to attenuate the instabilities. Combustion instabilities are a major concern in ground-based gas turbine engines because of NOx emissions. The tendency is to run lean, an equivalence ratio less than 1, to reduce the combustion temperature and thus reduce the NOx emissions; however, running the combustion lean makes it very susceptible to combustion instability.

The Rayleigh Criterion is the basis for analysis of thermoacoustic combustion instability and is evaluated using the Rayleigh Index over one cycle of instability

$$G(x) = \frac{1}{T}\int_T q'(x,t)p'(x,t)dt$$

where q' is the heat release rate perturbation and p' is the pressure fluctuation. When the heat release oscillations are in phase with the pressure oscillations, the Rayleigh Index is positive and the magnitude of the thermo acoustic instability is maximised.

On the other hand, if the Rayleigh Index is negative, then thermoacoustic damping occurs. The Rayleigh Criterion implies that a thermoacoustic instability can be optimally controlled by having heat release oscillations 180 degrees out of phase with pressure oscillations at the same frequency. This minimizes the Rayleigh Index.

Combustion Triangle

Essential conditions for combustion to occur:

- Presence of fuel.
- Presence of oxidizer (Not essentially oxygen).
- They must be in right proportions.
- The proportion will be dictated by flammability limit.
- Ignition energy.

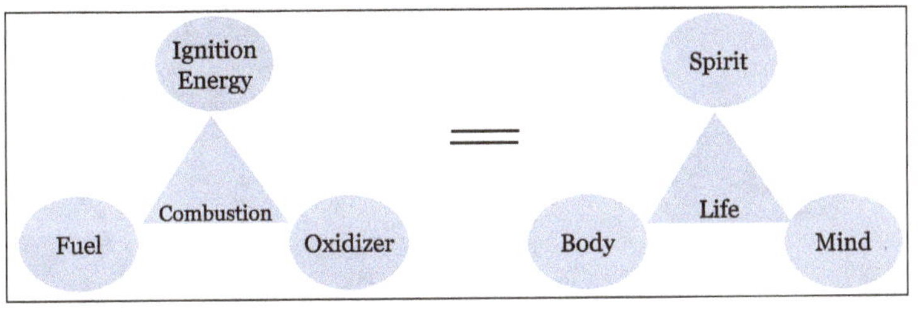

Industrial Process

- Thermal energy for process chemical plants, sugar industries, food processing industries are obtained through combustion.
- Iron, steel and other metals are produced from raw materials through combustion.
- Heat treatment and annealing of metals.
- Rotary kilns are used to produce Portlant cement.

Sugar Industry

Food Processing

Process Chemical Plant.

Steel Plant.

Transportation

- Surface transport vehicles are operated by reciprocating IC Engines.
- Gas turbine combustors are used widely in air and marine transportation sectors.

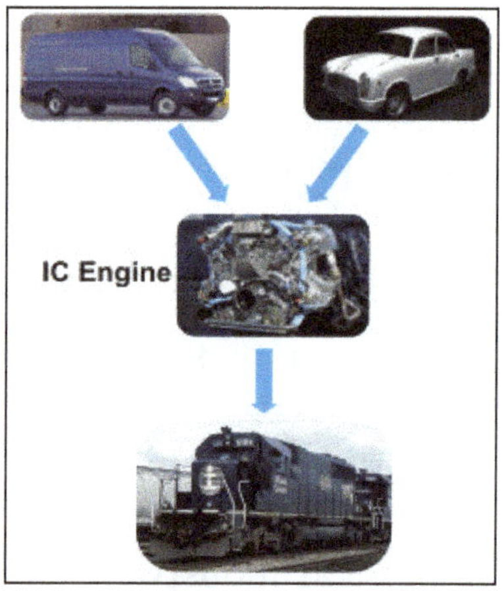

Power Generation

- Most of the thermal power plants are operated by burning coal.
- Recently gas turbine power plants are coming up.

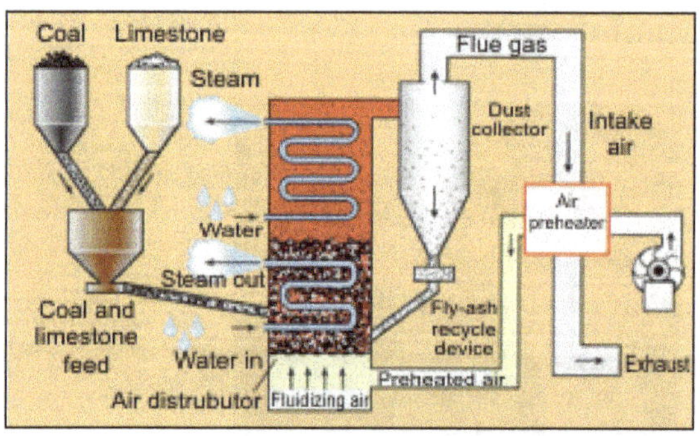

Fluidized Bed Power Plant.

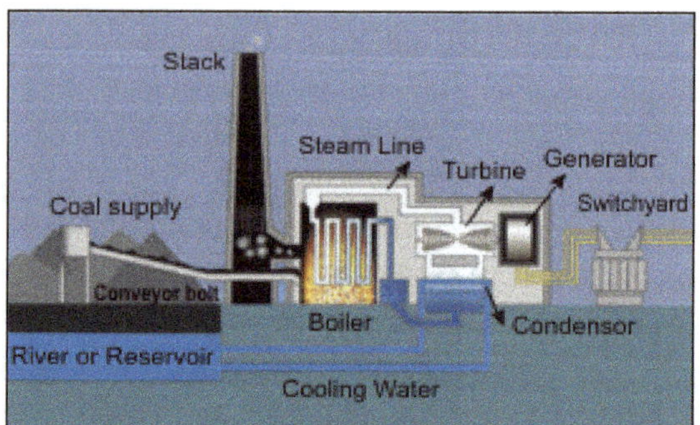

Coal Power Plant.

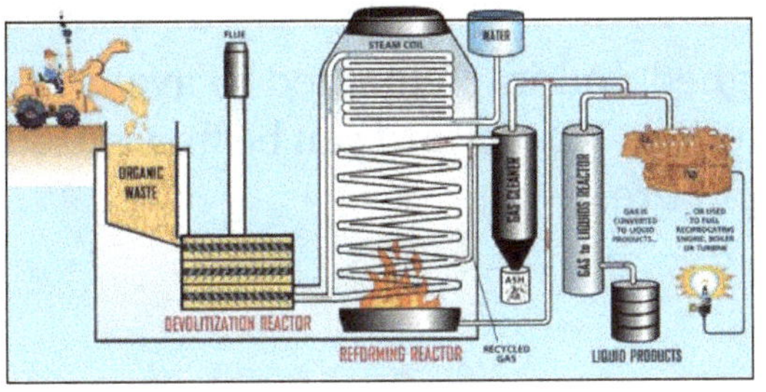

Biomass Based Power Plant.

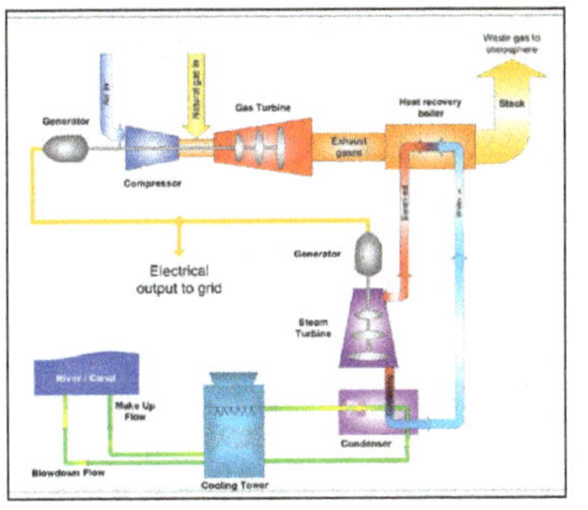

Gas Turbine Power Plant.

Waste Disposal

- Combustion finds application in disposing waste materials.
- Incinerators are used to dispose domestic and industrial wastes.
- In modern hospitals, incinerators are used to dispose hospital wastes safely.

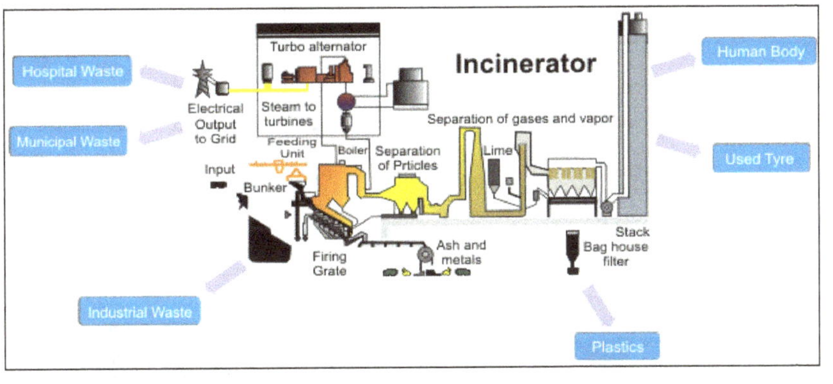

Fire

- Sometimes fire causes damage to life and property.
- Forest Fire: Damages natural resources and lives.

- Structural Fire: Damages property and human lives.
- Effective fire breakers should be designed and implemented to avoid fire hazard. By using proper construction materials, Fire hazard can be minimized. Marine life is very much affected by oil spill fire.

Environmental Pollution

- Combustion of any fuel produces certain amount of emissions such as smoke, ash, soot, and other harmful gases.
- Major pollution generated in combustion system are CO, CO_2, NO, NO_2, SO_2, ash, etc.
- These are due to incomplete combustion and can be minimized by increasing the residence time of fuel-oxidizer mixture in the combustor.

Transportation	7	79	60	15	6
Stationary Combustion Systems/Electricity	8	<1	<1	2	69
Industrial Process	23	8	32	13	25
Miscellaneous	62	13	8	70	<1

Fuel

Wood was one of the first fuels to be used by humans and is still the primary energy source in much of the world.

A fuel is any material that can be made to react with other substances so that it releases chemical or nuclear energy as heat or to be used for work. The concept was originally applied solely to those materials capable of releasing chemical energy but has since also been applied to other sources of heat energy such as nuclear energy (via nuclear fission and nuclear fusion).

The heat energy released by reactions of fuels is converted into mechanical energy via a heat engine. Other times the heat itself is valued for warmth, cooking, or industrial processes, as well as the illumination that comes with combustion. Fuels are also used in the cells of organisms in a process known as cellular respiration, where organic molecules are oxidized to release usable energy. Hydrocarbons and related oxygen-containing molecules are by far the most common source of fuel used by humans, but other substances, including radioactive metals, are also utilized.

Fuels are contrasted with other substances or devices storing potential energy, such as those that directly release electrical energy (such as batteries and capacitors) or mechanical energy (such as flywheels, springs, compressed air, or water in a reservoir).

History

The first known use of fuel was the combustion of wood or sticks by *Homo erectus* near 2,000,000 (two million) years ago. Throughout most of human history fuels derived from plants or animal fat were only used by humans. Charcoal, a wood derivative, has been used since at least 6,000 BCE for melting metals. It was only supplanted by coke, derived from coal, as European forests started to become depleted around the 18th century. Charcoal briquettes are now commonly used as a fuel for barbecue cooking.

Coal was first used as a fuel around 1000 BCE in China. With the energy in the form of chemical energy that could be released through combustion, but the concept development of the steam engine in the United Kingdom in 1769, coal came into more common use as a power source. Coal was later used to drive ships and locomotives. By the 19th century, gas extracted from coal was being used for street lighting in London. In the 20th and 21st centuries, the primary use of coal is to generate electricity, providing 40% of the world's electrical power supply in 2005.

Fossil fuels were rapidly adopted during the industrial revolution, because they were more concentrated and flexible than traditional energy sources, such as water power. They have become a pivotal part of our contemporary society, with most countries in the world burning fossil fuels in order to produce power.

Currently the trend has been towards renewable fuels, such as biofuels like alcohols.

Chemical

Chemical fuels are substances that release energy by reacting with substances around them, most notably by the process of combustion. Most of the chemical energy released in combustion was not stored in the chemical bonds of the fuel, but in the weak double bond of molecular oxygen.

Chemical fuels are divided in two ways. First, by their physical properties, as a solid, liquid or gas. Secondly, on the basis of their occurrence: *primary (natural fuel)* and *secondary (artificial fuel)*. Thus, a general classification of chemical fuels is:

General types of chemical fuels		
	Primary (natural)	**Secondary (artificial)**
Solid fuels	wood, coal, peat, dung, etc.	coke, charcoal
Liquid fuels	petroleum	diesel, gasoline, kerosene, LPG, coal tar, naphtha, ethanol
Gaseous fuels	natural gas	hydrogen, propane, methane, coal gas, water gas, blast furnace gas, coke oven gas, CNG

Solid Fuel

Coal is an important solid fuel.

Solid fuel refers to various types of solid material that are used as fuel to produce energy and provide heating, usually released through combustion. Solid fuels include wood, charcoal, peat, coal, Hexamine fuel tablets, and pellets made from wood, corn, wheat, rye and other grains. Solid-fuel rocket technology also uses solid fuel. Solid fuels have been used by humanity for many years to create fire. Coal was the fuel source which enabled the industrial revolution, from firing furnaces, to running steam engines. Wood was also extensively used to run steam locomotives. Both peat and coal are still used in electricity generation today. The use of some solid fuels (e.g. coal) is restricted or prohibited in some urban areas, due to unsafe levels of toxic emissions. The use of other solid fuels such as wood is decreasing as heating technology and the availability of good quality fuel improves. In some areas, smokeless coal is often the only solid fuel used. In Ireland, peat briquettes are used as smokeless fuel. They are also used to start a coal fire.

Liquid Fuels

Liquid fuels are combustible or energy-generating molecules that can be harnessed to create mechanical energy, usually producing kinetic energy; they also must take the shape of their container. It is the fumes of liquid fuels that are flammable instead of the fluid.

A gasoline station.

Most liquid fuels in widespread use are derived from the fossilized remains of dead plants and animals by exposure to heat and pressure in the Earth's crust. However, there are several types, such as hydrogen fuel (for automotive uses), ethanol, jet fuel and biodiesel which are all categorized as a liquid fuel. Emulsified fuels of oil-in-water such as orimulsion have been developed a way to make heavy oil fractions usable as liquid fuels. Many liquid fuels play a primary role in transportation and the economy.

Some common properties of liquid fuels are that they are easy to transport, and can be handled with relative ease. Also they are relatively easy to use for all engineering applications, and home use. Fuels like kerosene are rationed in some countries, for example available in government subsidized shops in India for home use.

Conventional diesel is similar to gasoline in that it is a mixture of aliphatic hydrocarbons extracted from petroleum. Kerosene is used in kerosene lamps and as a fuel for cooking, heating, and small engines. Natural gas, composed chiefly of methane, can only exist as a liquid at very low temperatures (regardless of pressure), which limits its direct use as a liquid fuel in most applications. LP gas is a mixture of propane and butane, both of which are easily compressible gases under standard atmospheric conditions. It offers many of the advantages of compressed natural gas (CNG), but is denser than air, does not burn as cleanly, and is much more easily compressed. Commonly used for cooking and space heating, LP gas and compressed propane are seeing increased use in motorized vehicles; propane is the third most commonly used motor fuel globally.

Gaseous Fuels

Fuel gas is any one of a number of fuels that under ordinary conditions are gaseous. Many fuel gases are composed of hydrocarbons (such as methane or propane), hydrogen, carbon monoxide, or mixtures thereof. Such gases are sources of potential heat energy or light energy that can be readily transmitted and distributed through pipes from the point of origin directly to the place of consumption. Fuel gas is contrasted with liquid fuels and from solid fuels, though some fuel gases are liquefied for storage or transport. While their gaseous nature can be advantageous, avoiding the difficulty of transporting solid fuel and the dangers of spillage inherent in liquid fuels, it can also be dangerous. It is possible for a fuel gas to be undetected and collect in certain areas, leading to the risk of a gas explosion. For this reason, odorizers are added to most fuel gases so that they may be detected by a distinct smell. The most common type of fuel gas in current use is natural gas.

A 20-pound (9.1 kg) propane cylinder.

Biofuels

Biofuel can be broadly defined as solid, liquid, or gas fuel consisting of, or derived from biomass. Biomass can also be used directly for heating or power—known as *biomass fuel*. Biofuel can be produced from any carbon source that can be replenished rapidly e.g. plants. Many different plants and plant-derived materials are used for biofuel manufacture.

Perhaps the earliest fuel employed by humans is wood. Evidence shows controlled fire was used up to 1.5 million years ago at Swartkrans, South Africa. It is unknown which hominid species first used fire, as both *Australopithecus* and an early species of *Homo* were present at the sites. As a fuel, wood has remained in use up until the present day, although it has been superseded for many purposes by other sources. Wood has an energy density of 10–20 MJ/kg.

Recently biofuels have been developed for use in automotive transport (for example Bioethanol and Biodiesel), but there is widespread public debate about how carbon efficient these fuels are.

Fossil Fuels

Extraction of petroleum.

Fossil fuels is hydrocarbons, primarily coal and petroleum (liquid petroleum or natural gas), formed from the fossilized remains of ancient plants and animals by exposure to high heat and pressure in the absence of oxygen in the Earth's crust over hundreds of millions of years. Commonly, the term fossil fuel also includes hydrocarbon-containing natural resources that are not derived entirely from biological sources, such as tar sands. These latter sources are properly known as *mineral fuels*.

Fossil fuels contain high percentages of carbon and include coal, petroleum, and natural gas. They range from volatile materials with low carbon:hydrogen ratios like methane, to liquid petroleum to nonvolatile materials composed of almost pure carbon, like anthracite coal. Methane can be found in hydrocarbon fields, alone, associated with oil, or in the form of methane clathrates. Fossil fuels formed from the fossilized remains of dead plants by exposure to heat and pressure in the Earth's crust over millions of years. This biogenic theory was first introduced by German scholar Georg Agricola in 1556 and later by Mikhail Lomonosov in the 18th century.

It was estimated by the Energy Information Administration that in 2007 primary sources of energy consisted of petroleum 36.0%, coal 27.4%, natural gas 23.0%, amounting to an 86.4% share for fossil fuels in primary energy consumption in the world. Non-fossil sources in 2006 included hydroelectric 6.3%, nuclear 8.5%, and others (geothermal, solar, tidal, wind, wood, waste) amounting to 0.9%. World energy consumption was growing about 2.3% per year.

Fossil fuels are non-renewable resources because they take millions of years to form, and reserves are being depleted much faster than new ones are being made. So we must conserve these fuels and use them judiciously. The production and use of fossil fuels raise environmental concerns. A global movement toward the generation of renewable energy is therefore under way to help meet increased energy needs. The burning of fossil fuels produces around 21.3 billion tonnes (21.3 gigatonnes) of carbon dioxide (CO_2) per year, but it is estimated that natural processes can only absorb about half of that amount, so there is a net increase of 10.65 billion tonnes of atmospheric carbon dioxide per year (one tonne of atmospheric carbon is equivalent to 44/12 or 3.7 tonnes of carbon dioxide). Carbon dioxide is one of the greenhouse gases that enhances radiative forcing and contributes to global warming, causing the average surface temperature of the Earth to rise in response, which the vast majority of climate scientists agree will cause major adverse effects. Fuels are a source of energy.

Energy

The amount of energy from different types of fuel depends on the stoichiometric ratio, the chemically correct air and fuel ratio to ensure complete combustion of fuel, and its Specific energy, the energy per unit mass.

Energy Capacities of Common Types of Fuel				
Fuel	**Specific energy (MJ/kg)**	**AFR stoich.**	**FAR stoich.**	**Energy @ λ=1 ($MJ/kg_{(Air)}$)**
Diesel	48	14.5 : 1	0.069 : 1	3.310
Ethanol	26.4	9 : 1	0.111 : 1	2.933
Gasoline	46.4	14.7 : 1	0.068 : 1	3.156
Hydrogen	142	34.3 : 1	0.029 : 1	4.140
Kerosene	46	15.6 : 1	0.064 : 1	2.949
LPG	46.4	17.2 : 1	0.058 : 1	2.698
Methanol	19.7	6.47 : 1	0.155 : 1	3.045
Nitromethane	11.63	1.7 : 1	0.588 : 1	6.841

1 MJ ≈ 0.28 kWh ≈ 0.37 HPh.

Nuclear

CANDU fuel bundles Two CANDU ("CANada Deuterium Uranium") fuel bundles, each about 50 cm long and 10 cm in diameter.

Nuclear fuel is any material that is consumed to derive nuclear energy. Technically speaking, All matter can be a nuclear fuel because any element under the right conditions will release nuclear energy, but the materials commonly referred to as nuclear fuels are those that will produce energy without being placed under extreme duress. Nuclear fuel is a material that can be 'burned' by nuclear fission or fusion to derive nuclear energy. *Nuclear fuel* can refer to the fuel itself, or to physical objects (for example bundles composed of fuel rods) composed of the fuel material, mixed with structural, neutron moderating, or neutron reflecting materials.

Most nuclear fuels contain heavy fissile elements that are capable of nuclear fission. When these fuels are struck by neutrons, they are in turn capable of emitting neutrons when they break apart. This makes possible a self-sustaining chain reaction that releases energy with a controlled rate in a nuclear reactor or with a very rapid uncontrolled rate in a nuclear weapon.

The most common fissile nuclear fuels are uranium-235 (^{235}U) and plutonium-239 (^{239}Pu). The actions of mining, refining, purifying, using, and ultimately disposing of nuclear fuel together make up the nuclear fuel cycle. Not all types of nuclear fuels create power from nuclear fission. Plutonium-238 and some other elements are used to produce small amounts of nuclear power by radioactive decay in radioisotope thermoelectric generators and other types of atomic batteries. Also, light nuclides such as tritium (^{3}H) can be used as fuel for nuclear fusion. Nuclear fuel has the highest energy density of all practical fuel sources.

Fission

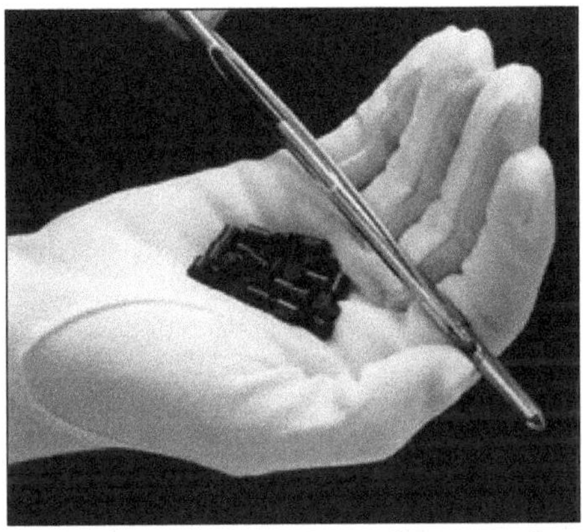

Nuclear fuel pellets are used to release nuclear energy.

The most common type of nuclear fuel used by humans is heavy fissile elements that can be made to undergo nuclear fission chain reactions in a nuclear fission reactor; *nuclear fuel* can refer to the material or to physical objects (for example fuel bundles composed of fuel rods) composed of the fuel material, perhaps mixed with structural, neutron moderating, or neutron reflecting materials. The most common fissile nuclear fuels are U and Pu, and the actions of mining, refining, purifying, using, and ultimately disposing of these elements together make up the nuclear fuel cycle, which is important for its relevance to nuclear power generation and nuclear weapons.

Fusion

Fuels that produce energy by the process of nuclear fusion are currently not utilized by humans but are the main source of fuel for stars. Fusion fuels tend to be light elements such as hydrogen which will combine easily. Energy is required to start fusion by raising temperature so high all materials would turn into plasma, and allow nuclei to collide and stick together with each other before repelling due to electric charge. This process is called fusion and it can give out energy.

In stars that undergo nuclear fusion, fuel consists of atomic nuclei that can release energy by the absorption of a proton or neutron. In most stars the fuel is provided by hydrogen, which can combine to form helium through the proton-proton chain reaction or by the CNO cycle. When the hydrogen fuel is exhausted, nuclear fusion can continue with progressively heavier elements, although the net energy released is lower because of the smaller difference in nuclear binding energy. Once iron-56 or nickel-56 nuclei are produced, no further energy can be obtained by nuclear fusion as these have the highest nuclear binding energies. The elements then on use up energy instead of giving off energy when fused. Therefore, fusion stops and the star dies. In attempts by humans, fusion is only carried out with hydrogen (isotope of 2 and 3) to form helium-4 as this reaction gives out the most net energy. Electric confinement (ITER), inertial confinement(heating by laser) and heating by strong electric currents are the popular methods used.

World Trade

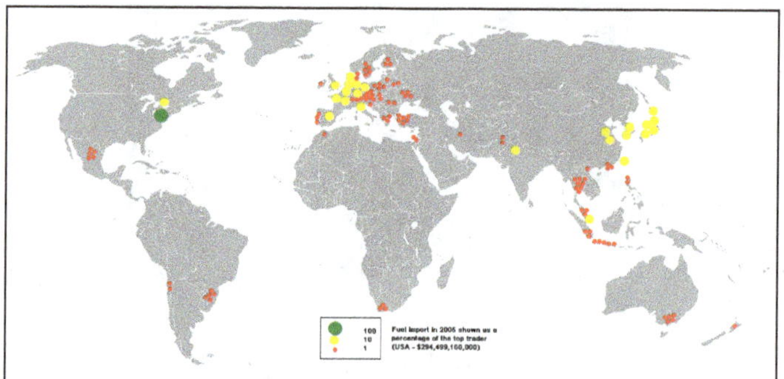

Fuel imports in 2005

The World Bank reported that the USA was the top fuel importer in 2005 followed by the EU and Japan.

Solid Fuels and Oxidizers

Solid Fuels

- Wood
- Coal
- Charcoal
- Soft Coke
- Biomass
- Animal dung

Constituents of Solid Fuel

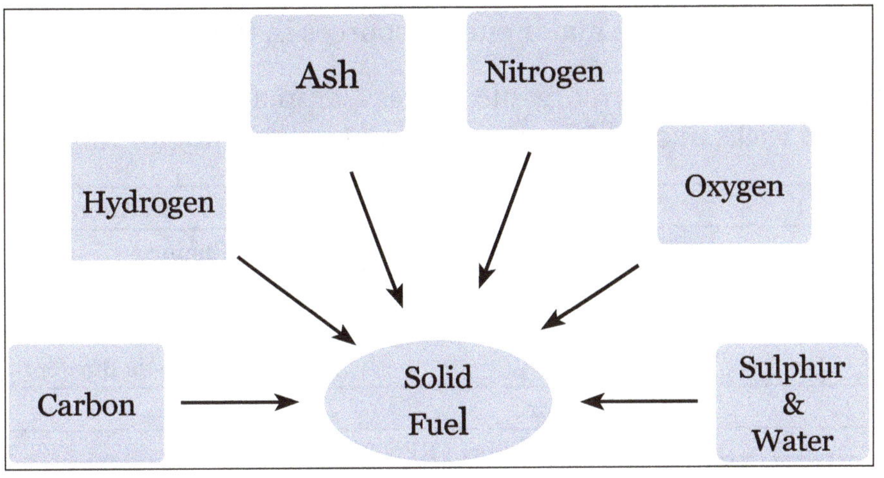

Types of Solid Fuels and Oxidizers

1	Biomass (Wood, Saw Dust, Rice Husk, Rice Straw, Wheat Straw, etc)	Air/O_2	Domestic Burner, Engine With Producer Gas
2	Coal, Coke, Charcoal	do	do
3	Special Fuels Nitrocellulose (NC), HTPB, CTPB	Nitroglycerine, Ammonium Perchlorate , Ammonium Nitrate, Nitrogen Tetraoxide	Solid Propellant Rocket, Hybrid Rocket

Oxygen, Water and Ash Content of Certain Solid Fuels

Moisture in Solid Fuel

1. Free.

2. Bound water.

Fuel moisture content will affect rate of combustion and overall efficiency.

Ash: The inorganic materials, which remain as residue even after complete combustion.

Ash content affects the performance of the combustion system.

Ash content causes fouling of the boilers.

Wood	40-45%	15-70%	0.1-1.0%
Peat	30-35%	70-90%	0.1-20%
Lignite coal	20-25%	20-30%	>5%
Bituminous coal	3-5%	10-5%	>5%
Anthracite coal	1-2%	2-4%	>5%

Liquid Fuels and Oxidizers

- Liquid fuel is one of the major energy sources in the transport sector.
- Crude oil is formed from organic sources, animals, vegetables – which are entrapped in rocks under high pressure and temperature for million years.

1	Gasoline	Air	S.I. Engine, Aircraft Piston Engine
2	HSD	Air	C.I. Engine
3	Furnace Oil	Air	Furnaces
4	Kerosene	Air	Aircraft, Gas Turbine, Engines Ramjet, Domestic Burner
5	Alcohols	Air	I.C. Engine
6	Hydrazine, UDMH, MMH, Liquid Hydrogen, Triethyl Amine	Liquid O_2, RFNA (Red Fuming Nitric Acid) N_2O_4	Liquid propellant rocket Engines
7	Hydrogen, Kerocene	Air	Ramjet/Scramjet

Refinery End-products of Typical Crude Oil

- Crude oil undergoes several process in the refinery.
- Generally separation of petroleum constituents occur in the distillation column.
- Constituents of typical crude oil is shown below.

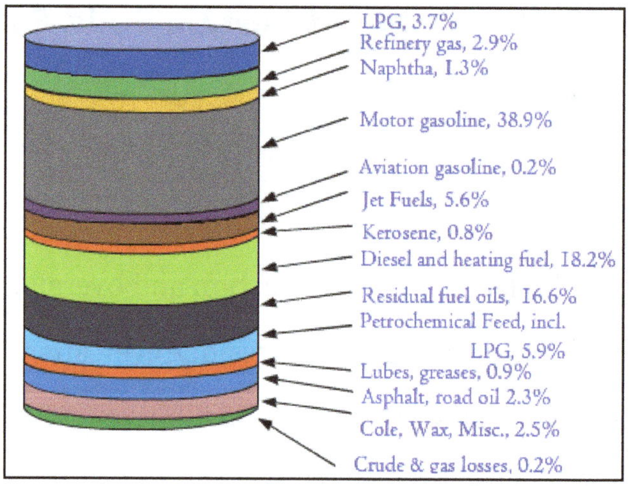

Bomb Calorimeter

- Used to determine the calorific value of the liquid fuel.
- Liquid is burnt in the bomb in the presence of oxygen at about 2.5 MPa.
- The change in temperature in the water bath provides the calorific value of the fuel.

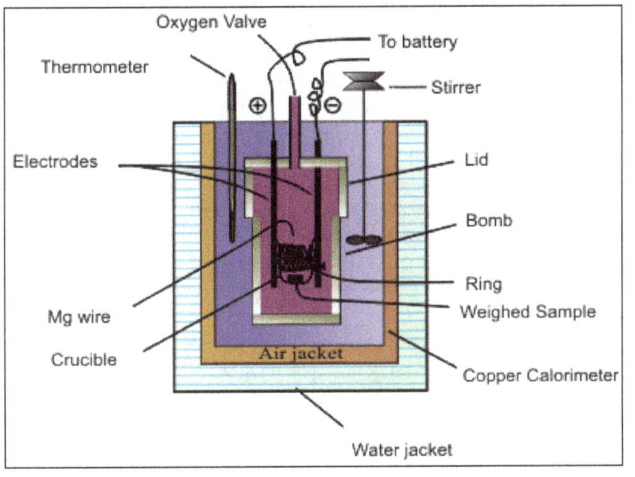

Properties of Liquid Fuels

Specific Gravity: Ratio of mass density of fuel to mass density of water at the same temperature:

$$SG = \frac{\rho_{fuel}}{\rho_{water}} (at\ the\ same\ temperature)$$

Reference temperature for fuel and water: 288.8 K

American Petroleum Institute (API) Scale:

$$APISG = \frac{141.5}{SG} - 131.4$$

Relation between APISG and HHV:

For Gasoline: $\mathrm{HHV = LHV + 93\ (APISG - 10) kJ / kg}$

For Kerosene: $\mathrm{HHV = LHV + 93\ (APISG - 10) kJ / kg}$

Auto Ignition Temperature : The lowest temperature required to make the combustion self sustained without any external aid.

Flash Point: Minimum temperature at which liquid fuel will produce sufficient vapors to form a flammable mixture with air. Indicates maximum temperature at which liquid fuel can be stored without any fire hazard.

Fire Point: Minimum temperature at which liquid fuel produces sufficient vapors to form a flammable mixture with air that continuously supports combustion establishing flame instead of just flashing.

Smoke Point: Measure of the tendency of a liquid fuel to produce soot.

Properties of Common Liquid Fuels

Specific gravity	0.72 - 0.78	0.85	0.796	0.82	0.71
Kinematics viscosity @ 293 K (m^2/s)	0.8 X 10^{-6}	2.5 X 10^{-6}	0.75 X 10^{-6}	3.626 X 10^{-6}	--
Boiling point range (K) @ STP	303 - 576	483 - 508	338	423-473	442
Flash point (K)	230	325	284	311	325
Auto ignition temperature (K)	643	527	737	483	--
Stoichiometric air/fuel by weight	14.7	14.7	6.45	15	15.1
Heat of Vaporization (kJ/kg)	380	375	1185	298.5	--
Lower heating value (MJ/kg)	43.5	45	20.1	45.2	43.3

Types of Fuels and Oxidizers

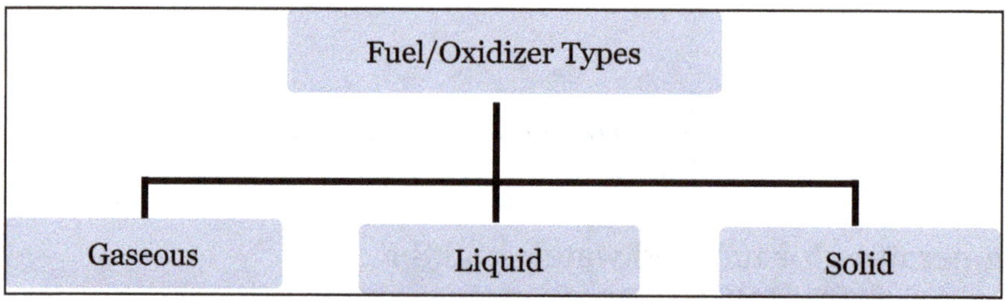

Gaseous Fuel and Oxidizer

Why gaseous fuels are preferred over liquid and solid fuels?

- Easier to control emissions.
- Easier to burn for higher efficiency.
- Gas handling system is less expensive.
- Commonly used gaseous fuels: CNG, LPG, Biogas, Producer gas, Coke oven gas, Acetylene, Methane, Hydrogen and Propane.

Types of Gaseous Fuel and Oxidizer

1	LPG	Air/O_2	Domestic Burner, Furnace
2	Natural Gas (NG)	Air/O_2	IC Engines, Furnaces
3	Producer Gas	Air/O_2	EC/IC Engines
4	CH_4, C_3H_8, H_2	Air/O_2	EC/IC Engines
5	Biogas	Air/O_2	EC/IC Engines, Burners
6	Acetylene	Air/O_2	Gas welding, Gas cutting
			* *EC=External Combustion*
			IC =Internal Combustion

Composition of Some Gaseous Fuels

Fuel	CO_2	O_2	N_2	CO	H_2	CH_4	C_2H_6	C_3H_8	C_4H_{10}
LPG	-	-	-	-	-	-	-	70	30
Natural Gas	-	-	5	-	-	90	5	-	-
Producer Gas	8	0.1	50	23.2	17.7	1	-	-	-
Propane	-	-	-	-	-	-	2.2	97.3	0.5
Biogas	33	-	1	-	1	65	-	-	-

Characterization of a Gaseous Fuel

Heating Value:

- Amount of heat released per unit volume when it undergoes oxidation at normal pressure and temperature (0.1 MPa and 298 K).
- Lower heating value (LHV) – amount of heat released by burning 1 kg of fuel assuming the latent heat of vaporization in the reaction products is not recovered.

$$LHV = HHV - \frac{m_{H_2O}}{m_{fuel}} \Delta H_v$$

- Higher heating value (HHV) – heating value of the fuel when water is condensed.
- ΔH_v is the Latent heat of vaporization of water at 298.15 K

Junker's Calorimeter

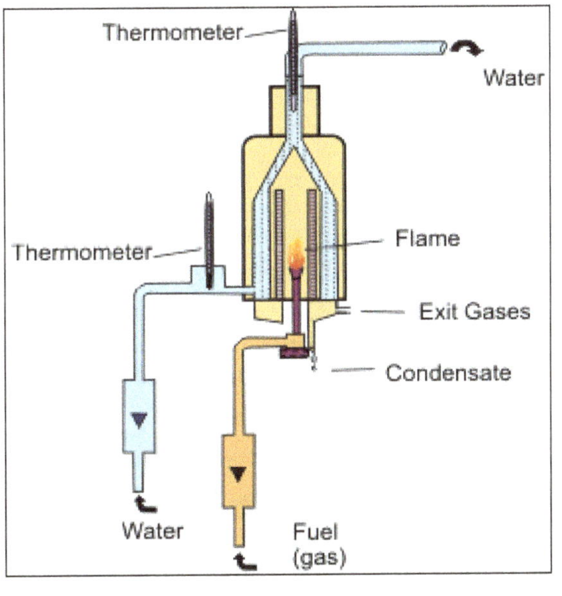

- Determines the heating value of the gaseous fuel.
- Fuel and air are burnt in a burner.
- Cooling water in the water jacket-absorbs the heat released during combustion.
- Heating value- calculated from the water flow rate and rise in temperature.

Thermodynamics of Combustion

Thermodynamic Properties

Thermodynamic Properties

Intensive	Extensive
Pressure (P)	Mass (m)
Temperature (T)	Volume (V)
Sp. Entropy (s)	Enthalpy (H)

Intensive	• Independent of mass • Extensive property/mass
Extensive	• Dependent of mass • Denoted by capital letters

Ideal Gas Law

$$PV=nR_uT$$

Where, P= Pressure
T= Temperature
n= Number of moles of the gas
R_u= Universal Gas Constant= 8.314 kJ/kmole K

Valid only when:

- No intermolecular force between molecules
- Molecules occupy negligible volume

Gas Mixture

Dalton's Law	Total pressure of a gaseous mixture is the sum of the pressure which each component would exert if it alone occupies the same volume and temperature of mixture.
Gibb's Theorem	Internal energy of a mixture of ideal gases is equal to the mole/mass fraction weighed sums of the internal energy of individual components of mixture.

Gibbs-Dalton law: Employed to extract properties of mixture from individual gases.
Gibb's Theorem: Specific molar internal energy of the mixture from the constituent species.

Entropy of Mixing

In thermodynamics the entropy of mixing is the increase in the total entropy when several initially separate systems of different composition, each in a thermodynamic state of internal equilibrium, are mixed without chemical reaction by the thermodynamic operation of removal of impermeable partition(s) between them, followed by a time for establishment of a new thermodynamic state of internal equilibrium in the new unpartitioned closed system.

In general, the mixing may be constrained to occur under various prescribed conditions. In the customarily prescribed conditions, the materials are each initially at a common temperature and pressure, and the new system may change its volume, while being maintained at that same constant temperature, pressure, and chemical component masses. The volume available for each material to explore is increased, from that of its initially separate compartment, to the total common final volume. The final volume need not be the sum of the initially separate volumes, so that work can be done on or by the new closed system during the process of mixing, as well as heat being transferred to or from the surroundings, because of the maintenance of constant pressure and temperature.

The internal energy of the new closed system is equal to the sum of the internal energies of the initially separate systems. The reference values for the internal energies should be specified in a way that is constrained to make this so, maintaining also that the internal energies are respectively proportional to the masses of the systems.

For concision in this section, the term 'ideal material' is used to refer to an ideal gas (mixture) or an ideal solution.

In a process of mixing of ideal materials, the final common volume is the sum of the initial separate compartment volumes. There is no heat transfer and no work is done. The entropy of mixing is entirely accounted for by the diffusive expansion of each material into a final volume not initially accessible to it.

On the mixing of non-ideal materials, the total final common volume may be different from the sum of the separate initial volumes, and there may occur transfer of work or heat, to or from the surroundings; also there may be a departure of the entropy of mixing from that of the corresponding ideal case. That departure is the main reason for interest in entropy of mixing. These energy and entropy variables and their temperature dependences provide valuable information about the properties of the materials.

On a molecular level, the entropy of mixing is of interest because it is a macroscopic variable that provides information about constitutive molecular properties. In ideal materials, intermolecular forces are the same between every pair of molecular kinds, so that a molecule feels no difference between other molecules of its own kind and of those of the other kind. In non-ideal materials, there may be differences of intermolecular forces or specific molecular effects between different species, even though they are chemically non-reacting. The entropy of mixing provides information about constitutive differences of intermolecular forces or specific molecular effects in the materials.

The statistical concept of randomness is used for statistical mechanical explanation of the entropy of mixing. Mixing of ideal materials is regarded as random at a molecular level, and, correspondingly, mixing of non-ideal materials may be non-random.

Mixing of Ideal Materials at Constant Temperature and Pressure

In ideal materials, intermolecular forces are the same between every pair of molecular kinds, so that a molecule "feels" no difference between itself and its molecular neighbours. This is the reference case for examining corresponding mixings of non-ideal materials.

For example, two ideal gases, at the same temperature and pressure, are initially separated by a dividing partition.

Upon removal of the dividing partition, they expand into a final common volume (the sum of the two initial volumes), and the entropy of mixing $\Delta_{mix}S$ is given by,

$$\Delta_{mix}S = -nR(x_1 \ln x_1 + x_2 \ln x_2).$$

where R is the gas constant, n the total number of moles and x_i the mole fraction of component i, which initially occupies volume $V_i = x_i V$. After the removal of the partition, the $n_i = nx_i$ moles of component i may explore the combined volume V, which causes an entropy increase equal to $nx_i R \ln(V / V_i) = -nRx_i \ln x_i$ for each component gas.

In this case, the increase in entropy is due entirely to the irreversible processes of expansion of the two gases, and involves no heat or work flow between the system and its surroundings.

Gibbs Free Energy of Mixing

The Gibbs free energy change $\Delta_{mix}G = \Delta_{mix}H - T\Delta_{mix}S$ determines whether mixing at constant (absolute) temperature T and pressure p is a spontaneous process. This quantity combines two physical effects—the enthalpy of mixing, which is a measure of the energy change, and the entropy of mixing considered here.

For an ideal gas mixture or an ideal solution, there is no enthalpy of mixing ($\Delta_{mix}H$), so that the Gibbs free energy of mixing is given by the entropy term only:

$$\Delta_{mix}G = -T\Delta_{mix}S$$

For an ideal solution, the Gibbs free energy of mixing is always negative, meaning that mixing of ideal solutions is always spontaneous. The lowest value is when the mole fraction is 0.5 for a mixture of two components, or 1/n for a mixture of n components.

Solutions and Temperature Dependence of Miscibility

Ideal and Regular Solutions

The above equation for the entropy of mixing of ideal gases is valid also for certain liquid (or solid) solutions—those formed by completely random mixing so that the components move independently in the total volume. Such random mixing of solutions occurs if the interaction energies between unlike molecules are similar to the average interaction energies between like molecules. The value of the entropy corresponds exactly to random mixing for ideal solutions and for regular solutions, and approximately so for many real solutions.

For binary mixtures the entropy of random mixing can be considered as a function of the mole fraction of one component.

$$\Delta_{mix} S = -nR(x_1 \ln x_1 + x_2 \ln x_2) = -nR[x \ln x + (1-x)\ln(1-x)]$$

For all possible mixtures, $0 < x < 1$, so that $\ln x$ and $\ln(1-x)$ are both negative and the entropy of mixing $\Delta_{mix} S$ is positive and favors mixing of the pure components.

Also the curvature of $\Delta_{mix} S$ as a function of x is given by the second derivative,

$$\left(\frac{\partial^2 \Delta_{mix} S}{\partial x^2}\right)_{T,P} = -nR\left(\frac{1}{x} + \frac{1}{1-x}\right)$$

This curvature is negative for all possible mixtures $(0 < x < 1)$, so that mixing two solutions to form a solution of intermediate composition also increases the entropy of the system. Random mixing therefore always favors miscibility and opposes phase separation.

For ideal solutions, the enthalpy of mixing is zero so that the components are miscible in all proportions. For regular solutions a positive enthalpy of mixing may cause incomplete miscibility (phase separation for some compositions) at temperatures below the upper critical solution temperature (UCST). This is the minimum temperature at which the $-T\Delta_{mix} S$ term in the Gibbs energy of mixing is sufficient to produce miscibility in all proportions.

Systems with a Lower Critical Solution Temperature

Nonrandom mixing with a lower entropy of mixing can occur when the attractive interactions between unlike molecules are significantly stronger (or weaker) than the mean interactions between like molecules. For some systems this can lead to a lower critical solution temperature (LCST) or lower limiting temperature for phase separation.

For example, triethylamine and water are miscible in all proportions below 19°C, but above this critical temperature, solutions of certain compositions separate into two phases at equilibrium with each other. This means that $\Delta_{mix}G$ is negative for mixing of the two phases below 19°C and positive above this temperature. Therefore, $\Delta_{mix}S = -\left(\frac{\partial \Delta_{mix}G}{\partial T}\right)_P$ is negative for mixing of these two equilibrium phases. This is due to the formation of attractive hydrogen bonds between the two components that prevent random mixing. Triethylamine molecules cannot form hydrogen bonds with each other but only with water molecules, so in solution they remain associated to water molecules with loss of entropy. The mixing that occurs below 19°C is due not to entropy but to the enthalpy of formation of the hydrogen bonds.

Lower critical solution temperatures also occur in many polymer-solvent mixtures. For polar systems such as polyacrylic acid in 1,4-dioxane, this is often due to the formation of hydrogen bonds between polymer and solvent. For nonpolar systems such as polystyrene in cyclohexane, phase separation has been observed in sealed tubes (at high pressure) at temperatures approaching the liquid-vapor critical point of the solvent. At such temperatures the solvent expands much more rapidly than the polymer, whose segments are covalently linked. Mixing therefore requires contraction of the solvent for compatibility of the polymer, resulting in a loss of entropy.

Statistical Thermodynamical Explanation of the Entropy of Mixing of Ideal Gases

Since thermodynamic entropy can be related to statistical mechanics or to information theory, it is possible to calculate the entropy of mixing using these two approaches. Here we consider the simple case of mixing ideal gases.

Proof from Statistical Mechanics

Assume that the molecules of two different substances are approximately the same size, and regard space as subdivided into a square lattice whose cells are the size of the molecules. (In fact, any lattice would do, including close packing.) This is a crystal-like conceptual model to identify the molecular centers of mass. If the two phases are liquids, there is no spatial uncertainty in each one individually. (This is, of course, an approximation. Liquids have a “free volume”. This is why they are (usually) less dense than solids.) Everywhere we look in component 1, there is a molecule present, and likewise for component 2. After the two different substances are intermingled (assuming they are miscible), the liquid is still dense with molecules, but now there is uncertainty about what kind of molecule is in which location. Of course, any idea of identifying molecules in given locations is a thought experiment, not something one could do, but the calculation of the uncertainty is well-defined.

We can use Boltzmann's equation for the entropy change as applied to the mixing process,

$$\Delta_{mix} S = k_B \ln \Omega$$

where k_B is Boltzmann's constant. We then calculate the number of ways Ω of arranging N_1 molecules of component 1 and N_2 molecules of component 2 on a lattice, where,

$$N = N_1 + N_2$$

is the total number of molecules, and therefore the number of lattice sites. Calculating the number of permutations of objects, correcting for the fact that N_1 of them are *identical* to one another, and likewise for N_2,

$$\Omega = \mathrm{N}!/\,\mathrm{N}_1!\mathrm{N}_2!$$

After applying Stirling's approximation for the factorial of a large integer m:

$$\ln m! = \sum_k \ln k \approx \int_1^m dk \ln k = m \ln m - m,$$

the result is

$$\Delta_{mix} S = -k_B[N_1 \ln(N_1 / N) + N_2 \ln(N_2 / N)] = -k_B N[x_1 \ln x_1 + x_2 \ln x_2]$$

where we have introduced the mole fractions, which are also the probabilities of finding any particular component in a given lattice site.

$$x_1 = N_1 / N = p_1 \text{ and } x_2 = N_2 / N = p_2$$

Since the Boltzmann constant $k_B = R / N_A$, where N_A is Avogadro's number, and the number of molecules $N = nN_A$, we recover the thermodynamic expression for the mixing of two ideal gases,

$$\Delta_{mix} S = -nR[x_1 \ln x_1 + x_2 \ln x_2]$$

This expression can be generalized to a mixture of r components, N_i ,with $i = 1, 2, 3, \ldots, r$

$$\Delta_{mix} S = -k_B \sum_{i=1}^{r} N_i \ln(N_i / N) = -Nk_B \sum_{i=1}^{r} x_i \ln x_i = -nR \sum_{i=1}^{r} x_i \ln x_i$$

Relationship to Information Theory

The entropy of mixing is also proportional to the Shannon entropy or compositional uncertainty of information theory, which is defined without requiring Stirling's approximation. Claude Shannon introduced this expression for use in information theory, but similar formulas can be found as far back as the work of Ludwig Boltzmann and J. Willard Gibbs. The Shannon uncertainty is not the same as the Heisenberg uncertainty principle in quantum mechanics which is based on variance. The Shannon entropy is defined as:

$$H = -\sum_{i=1}^{r} p_i \ln(p_i)$$

where p_i is the probability that an information source will produce the i th symbol from an r-symbol alphabet and is independent of previous symbols. (thus i runs from 1 to r). H is then a measure of the expected amount of information ($\log p_i$) missing before the symbol is known or measured, or, alternatively, the expected amount of information supplied when the symbol becomes known. The set of messages of length N symbols from the source will then have an entropy of N*H.

The thermodynamic entropy is only due to positional uncertainty, so we may take the "alphabet" to be any of the r different species in the gas, and, at equilibrium, the probability that a given particle is of type i is simply the mole fraction x_i for that particle. Since we are dealing with ideal gases, the identity of nearby particles is irrelevant. Multiplying by the number of particles N yields the change in entropy of the entire system from the unmixed case in which all of the p_i were either 1 or 0. We again obtain the entropy of mixing on multiplying by the Boltzmann constant k_B .

$$\Delta_{mix} S = -Nk_B \sum_{i=1}^{r} x_i \ln x_i$$

So thermodynamic entropy with "r" chemical species with a total of N particles has a parallel to an information source that has "r" distinct symbols with messages that are N symbols long.

Application to Gases

In gases there is a lot more spatial uncertainty because most of their volume is merely empty space. We can regard the mixing process as allowing the contents of the two originally separate contents to expand into the combined volume of the two conjoined containers. The two lattices that allow us to conceptually localize molecular centers of mass also join. The total number of empty cells is the sum of the numbers of empty cells in the two components prior to mixing. Consequently, that part of the spatial uncertainty concerning whether *any* molecule is present in a lattice cell is the sum of the initial values, and does not increase upon "mixing".

Almost everywhere we look, we find empty lattice cells. Nevertheless, we do find molecules in a few occupied cells. When there is real mixing, for each of those few occupied cells, there is a contingent uncertainty about which kind of molecule it is. When there is no real mixing because the two substances are identical, there is no uncertainty about which kind of molecule it is. Using conditional probabilities, it turns out that the analytical problem for the small subset of occupied cells is exactly the same as for mixed liquids, and the *increase* in the entropy, or spatial uncertainty, has exactly the same form as obtained previously. Obviously the subset of occupied cells is not the same at different times. But only when there is real mixing and an occupied cell is found do we ask which kind of molecule is there.

Application to Solutions

If the solute is a crystalline solid, the argument is much the same. A crystal has no spatial uncertainty at all, except for crystallographic defects, and a (perfect) crystal allows us to localize the molecules using the crystal symmetry group. The fact that volumes do not add when dissolving a solid in a liquid is not important for condensed phases. If the solute is not crystalline, we can still use a spatial lattice, as good an approximation for an amorphous solid as it is for a liquid.

The Flory–Huggins solution theory provides the entropy of mixing for polymer solutions, in which the macromolecules are huge compared to the solvent molecules. In this case, the assumption is made that each monomer subunit in the polymer chain occupies a lattice site.

Note that solids in contact with each other also slowly interdiffuse, and solid mixtures of two or more components may be made at will (alloys, semiconductors, etc.). Again, the same equations for the entropy of mixing apply, but only for homogeneous, uniform phases.

Mixing Under Other Constraints

Mixing with and without Change of Available Volume

In the established customary usage, expressed in the lead section of this article, the entropy of mixing comes from two mechanisms, the intermingling and possible interactions of the distinct molecular species, and the change in the volume available for each molecular species, or the change in concentration of each molecular species. For ideal gases, the entropy of mixing at prescribed common temperature and pressure has nothing to do with mixing in the sense of intermingling and interactions of molecular species, but is only to do with expansion into the common volume.

According to Fowler and Guggenheim (1939/1965), the conflating of the just-mentioned two mechanisms for the entropy of mixing is well established in customary terminology, but can be confusing unless it is borne in mind that the independent variables are the common initial and final temperature and total pressure; if the respective partial pressures or the total volume are chosen as independent variables instead of the total pressure, the description is different.

- Mixing with Each Gas Kept at Constant Partial Volume, with Changing Total Volume:

In contrast to the established customary usage, "mixing" might be conducted reversibly at constant volume for each of two fixed masses of gases of equal volume, being mixed by gradually merging their initially separate volumes by use of two ideal semipermeable membranes, each permeable only to one of the respective gases, so that the respective volumes available to each gas remain constant during the merge. Either one of the common temperature or the common pressure is chosen to be independently controlled by the experimenter, the other being allowed to vary so as to maintain constant volume for each mass of gas. In this kind of "mixing", the final common volume is equal to each of the respective separate initial volumes, and each gas finally occupies the same volume as it did initially.

This constant volume kind of "mixing", in the special case of perfect gases, is referred to in what is sometimes called Gibbs' theorem. It states that the entropy of such "mixing" of perfect gases is zero.

- Mixing at Constant Total Volume and Changing Partial Volumes, with Mechanically Controlled Varying Pressure, and Constant Temperature:

An experimental demonstration may be considered. The two distinct gases, in a cylinder of constant total volume, are at first separated by two contiguous pistons made respectively of two suitably specific ideal semipermeable membranes. Ideally slowly and fictively reversibly, at constant temperature, the gases are allowed to mix in the volume between the separating membranes, forcing them apart, thereby supplying work to an external system. The energy for the work comes from the heat reservoir that keeps the temperature constant. Then, by externally forcing ideally slowly the separating membranes together, back to contiguity, work is done on the mixed gases, fictively reversibly separating them again, so that heat is returned to the heat reservoir at constant temperature. Because the mixing and separation are ideally slow and fictively reversible, the work supplied by the gases as they mix is equal to the work done in separating them again. Passing from fictive reversibility to physical reality, some amount of additional work, that remains external to the gases and the heat reservoir, must be provided from an external source for this cycle, as required by the second law of thermodynamics, because this cycle has only one heat reservoir at constant temperature, and the external provision of work cannot be completely efficient.

Gibbs' Paradox

- "Mixing" of Identical Species Versus Mixing of Closely Similar But non-identical Species.

For entropy of mixing to exist, the putatively mixing molecular species must be chemically or physically detectably distinct. Thus arises the so-called *Gibbs paradox*, as follows. If molecular species are identical, there is no entropy change on mixing them, because, defined in thermodynamic terms, there is no mass transfer, and thus no thermodynamically recognized process of mixing. Yet the slightest detectable difference in constitutive properties between the two species yields a thermodynamically recognized process of transfer with mixing, and a possibly considerable entropy change, namely the entropy of mixing.

The "paradox" arises because any detectable constitutive distinction, no matter how slight, can lead to a considerably large change in amount of entropy as a result of mixing. Though a continuous change in the properties of the materials that are mixed might make the degree of constitutive difference tend continuously to zero, the entropy change would nonetheless vanish discontinuously when the difference reached zero.

From a general physical viewpoint, this discontinuity is paradoxical. But from a specifically thermodynamic viewpoint, it is not paradoxical, because in that discipline the degree of constitutive difference is not questioned; it is either there or not there. Gibbs himself did not see it as paradoxical. Distinguishability of two materials is a constitutive, not a thermodynamic, difference, for the laws of thermodynamics are the same for every material, while their constitutive characteristics are diverse.

Though one might imagine a continuous decrease of the constitutive difference between any two chemical substances, physically it cannot be continuously decreased till it actually vanishes. It is hard to think of a smaller difference than that between ortho- and para-hydrogen. Yet they differ by a finite amount. The hypothesis, that the distinction might tend continuously to zero, is unphysical. This is neither examined nor explained by thermodynamics. Differences of constitution are explained by quantum mechanics, which postulates discontinuity of physical processes.

For a detectable distinction, some means should be physically available. One theoretical means would be through an ideal semi-permeable membrane. It should allow passage, backwards and forwards, of one species, while passage of the other is prevented entirely. The entirety of prevention should include perfect efficacy over a practically infinite time, in view of the nature of thermodynamic equilibrium. Even the slightest departure from ideality, as assessed over a finite time, would extend to utter non-ideality, as assessed over a practically infinite time. Such quantum phenomena as tunneling ensure that nature does not allow such membrane ideality as would support the theoretically demanded continuous decrease, to zero, of detectable distinction. The decrease to zero detectable distinction must be discontinuous.

For ideal gases, the entropy of mixing does not depend on the degree of difference between the distinct molecular species, but only on the fact that they are distinct; for non-ideal gases, the entropy of mixing can depend on the degree of difference of the distinct molecular species. The suggested or putative "mixing" of identical molecular species is not in thermodynamic terms a mixing at all, because thermodynamics refers to states specified by state variables, and does not permit an imaginary labelling of particles. Only if the molecular species are different is there mixing in the thermodynamic sense.

Thermodynamics

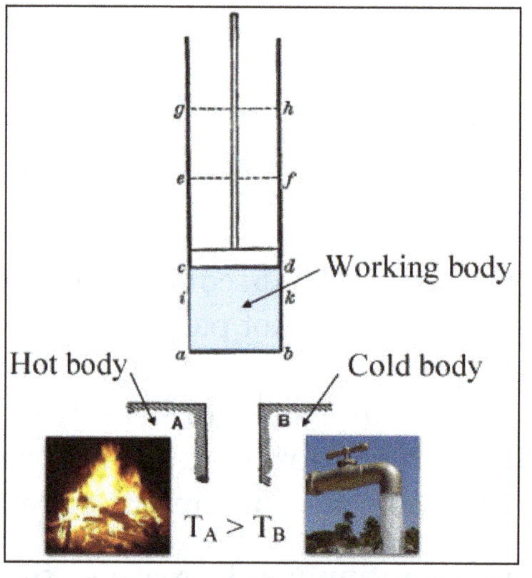

Annotated color version of the original 1824 Carnot heat engine showing the hot body (boiler), working body (system, steam), and cold body (water), the letters labeled according to the stopping points in Carnot cycle.

Thermodynamics is a branch of physics concerned with heat and temperature and their relation to energy and work. The behavior of these quantities is governed by the four laws of thermodynamics, irrespective of the composition or specific properties of the material or system in question. The laws of thermodynamics are explained in terms of microscopic constituents by statistical mechanics. Thermodynamics applies to a wide variety of topics in science and engineering, especially physical chemistry, chemical engineering and mechanical engineering.

Historically, thermodynamics developed out of a desire to increase the efficiency of early steam engines, particularly through the work of French physicist Nicolas Léonard Sadi Carnot (1824) who believed that engine efficiency was the key that could help France win the Napoleonic Wars. Scottish physicist Lord Kelvin was the first to formulate a concise definition of thermodynamics in 1854 which stated, *"Thermo-dynamics is the subject of the relation of heat to forces acting between contiguous parts of bodies, and the relation of heat to electrical agency."*

The initial application of thermodynamics to mechanical heat engines was extended early on to the study of chemical compounds and chemical reactions. Chemical thermodynamics studies the nature of the role of entropy in the process of chemical reactions and has provided the bulk of expansion and knowledge of the field. Other formulations of thermodynamics emerged in the following decades. Statistical thermodynamics, or statistical mechanics, concerned itself with statistical predictions of the collective motion of particles from their microscopic behavior. In 1909, Constantin Carathéodory presented a purely mathematical approach to the field in his axiomatic formulation of thermodynamics, a description often referred to as *geometrical thermodynamics*.

A description of any thermodynamic system employs the four laws of thermodynamics that form an axiomatic basis. The first law specifies that energy can be exchanged between physical systems as heat and work. The second law defines the existence of a quantity called entropy, that describes the direction, thermodynamically, that a system can evolve and quantifies the state of order of a system and that can be used to quantify the useful work that can be extracted from the system.

In thermodynamics, interactions between large ensembles of objects are studied and categorized. Central to this are the concepts of the thermodynamic *system* and its *surroundings*. A system is composed of particles, whose average motions define its properties, and those properties are in turn are related to one another through equations of state. Properties can be combined to express internal energy and thermodynamic potentials, which are useful for determining conditions for equilibrium and spontaneous processes.

The thermodynamicists representative of the original eight founding schools of thermodynamics. The schools with the most-lasting effect in founding the modern versions of thermodynamics are the Berlin school, particularly as established in Rudolf Clausius's 1865 textbook. *The Mechanical Theory of Heat*, the Vienna school, with the statistical mechanics of Ludwig Boltzmann, and the Gibbsian school at Yale University, American engineer Willard Gibbs' 1876 *On the Equilibrium of Heterogeneous Substances* launching chemical thermodynamics.

With these tools, thermodynamics can be used to describe how systems respond to changes in their environment. This can be applied to a wide variety of topics in science and engineering, such as engines, phase transitions, chemical reactions, transport phenomena, and even black holes. The results of thermodynamics are essential for other fields of physics and for chemistry, chemical engineering, aerospace engineering, mechanical engineering, cell biology, biomedical engineering, materials science, and economics, to name a few.

This section is focused mainly on classical thermodynamics which primarily studies systems in thermodynamic equilibrium. Non-equilibrium thermodynamics is often treated as an extension of the classical treatment, but statistical mechanics has brought many advances to that field.

History

The history of thermodynamics as a scientific discipline generally begins with Otto von Guericke who, in 1650, built and designed the world's first vacuum pump and demonstrated a vacuum using his Magdeburg hemispheres. Guericke was driven to make a vacuum in order to disprove Aristotle's long-held supposition that 'nature abhors a vacuum'. Shortly after Guericke, the English physicist and chemist Robert Boyle had learned of Guericke's designs and, in 1656, in coordination with English scientist Robert Hooke, built an air pump. Using this pump, Boyle and Hooke noticed a correlation between pressure, temperature, and volume. In time, Boyle's Law was formulated, which states that pressure and volume are inversely proportional. Then, in 1679, based on these concepts, an associate of Boyle's named Denis Papin built a steam digester, which was a closed vessel with a tightly fitting lid that confined steam until a high pressure was generated.

Later designs implemented a steam release valve that kept the machine from exploding. By watching the valve rhythmically move up and down, Papin conceived of the idea of a piston and a cylinder engine. He did not, however, follow through with his design. Nevertheless, in 1697, based on Papin's designs, engineer Thomas Savery built the first engine, followed by Thomas Newcomen in 1712. Although these early engines were crude and inefficient, they attracted the attention of the leading scientists of the time.

The fundamental concepts of heat capacity and latent heat, which were necessary for the development of thermodynamics, were developed by Professor Joseph Black at the University of Glasgow, where James Watt was employed as an instrument maker. Black and Watt performed experiments together, but it was Watt who conceived the idea of the external condenser which resulted in a large increase in steam engine efficiency. Drawing on all the previous work led Sadi Carnot, the "father of thermodynamics", to publish *Reflections on the Motive Power of Fire* (1824), a discourse on heat, power, energy and engine efficiency. The paper outlined the basic energetic relations between the Carnot engine, the Carnot cycle, and motive power. It marked the start of thermodynamics as a modern science.

The first thermodynamic textbook was written in 1859 by William Rankine, originally trained as a physicist and a civil and mechanical engineering professor at the University of Glasgow. The first and second laws of thermodynamics emerged simultaneously in the 1850s, primarily out of the works of William Rankine, Rudolf Clausius, and William Thomson (Lord Kelvin).

The foundations of statistical thermodynamics were set out by physicists such as James Clerk Maxwell, Ludwig Boltzmann, Max Planck, Rudolf Clausius and J. Willard Gibbs.

During the years 1873-76 the American mathematical physicist Josiah Willard Gibbs published a series of three papers, the most famous being *On the Equilibrium of Heterogeneous Substances*, in which he showed how thermodynamic processes, including chemical reactions, could be graphically analyzed, by studying the energy, entropy, volume, temperature and pressure of the thermodynamic system in such a manner, one can determine if a process would occur spontaneously. Also Pierre Duhem in the 19th century wrote about chemical thermodynamics. During the early 20th century, chemists such as Gilbert N. Lewis, Merle Randall, and E. A. Guggenheim applied the mathematical methods of Gibbs to the analysis of chemical processes.

Etymology

The etymology of *thermodynamics* has an intricate history. It was first spelled in a hyphenated form as an adjective (*thermo-dynamic*) and from 1854 to 1868 as the noun *thermo-dynamics* to represent the science of generalized heat engines.

American biophysicist Donald Haynie claims that *thermodynamics* was coined in 1840 from the Greek root *therme,* meaning heat and *dynamis,* meaning power. However, this etymology has been cited as unlikely.

Pierre Perrot claims that the term *thermodynamics* was coined by James Joule in 1858 to designate the science of relations between heat and power, however, Joule never used that term, but used instead the term *perfect thermo-dynamic engine* in reference to Thomson's 1849 phraseology.

By 1858, *thermo-dynamics,* as a functional term, was used in William Thomson's paper *An Account of Carnot's Theory of the Motive Power of Heat.*

Branches of Description

The study of thermodynamical systems has developed into several related branches, each using a different fundamental model as a theoretical or experimental basis, or applying the principles to varying types of systems.

Classical Thermodynamics

Classical thermodynamics is the description of the states of thermodynamic systems at near-equilibrium, that uses macroscopic, measurable properties. It is used to model exchanges of energy, work and heat based on the laws of thermodynamics. The qualifier *classical* reflects the fact that it represents the first level of understanding of the subject as it developed in the 19th century and describes the changes of a system in terms of macroscopic empirical (large scale, and measurable) parameters. A microscopic interpretation of these concepts was later provided by the development of *statistical mechanics*.

Statistical Mechanics

Statistical mechanics, also called statistical thermodynamics, emerged with the development of atomic and molecular theories in the late 19th century and early 20th century, and supplemented classical thermodynamics with an interpretation of the microscopic interactions between individual particles or quantum-mechanical states. This field relates the microscopic properties of individual atoms and molecules to the macroscopic, bulk properties of materials that can be observed on the human scale, thereby explaining classical thermodynamics as a natural result of statistics, classical mechanics, and quantum theory at the microscopic level.

Chemical Thermodynamics

Chemical thermodynamics is the study of the interrelation of energy with chemical reactions or with a physical change of state within the confines of the laws of thermodynamics.

Treatment of Equilibrium

Equilibrium thermodynamics is the systematic study of transformations of matter and energy in systems as they approach equilibrium. The word equilibrium implies a state of balance. In an equilibrium state there are no unbalanced potentials, or driving forces, within the system. A central aim in equilibrium thermodynamics is: given a system in a well-defined initial state, subject to accurately specified constraints, to calculate what the state of the system will be once it has reached equilibrium.

Non-equilibrium thermodynamics is a branch of thermodynamics that deals with systems that are not in thermodynamic equilibrium. Most systems found in nature are not in thermodynamic equilibrium because they are not in stationary states, and are continuously and discontinuously subject to flux of matter and energy to and from other systems. The thermodynamic study of non-equilibrium systems requires more general concepts than are dealt with by equilibrium thermodynamics. Many natural systems still today remain beyond the scope of currently known macroscopic thermodynamic methods.

Laws of Thermodynamics

Thermodynamics is principally based on a set of four laws which are universally valid when applied to systems that fall within the constraints implied by each. In the various theoretical descriptions of thermodynamics these laws may be expressed in seemingly differing forms, but the most prominent formulations are the following:

- Zeroth law of thermodynamics: *If two systems are each in thermal equilibrium with a third, they are also in thermal equilibrium with each other.*

This statement implies that thermal equilibrium is an equivalence relation on the set of thermodynamic systems under consideration. Systems are said to be in equilibrium if the small, random exchanges between them (e.g. Brownian motion) do not lead to a net change in energy. This law is tacitly assumed in every measurement of temperature. Thus, if one seeks to decide if two bodies are at the same temperature, it is not necessary to bring them into contact and measure any changes of their observable properties in time. The law provides an empirical definition of temperature and justification for the construction of practical thermometers.

The zeroth law was not initially recognized as a law, as its basis in thermodynamical equilibrium was implied in the other laws. The first, second, and third laws had been explicitly stated prior and found common acceptance in the physics community. Once the importance of the zeroth law for the definition of temperature was realized, it was impracticable to renumber the other laws, hence it was numbered the *zeroth law*.

- First law of thermodynamics: *The internal energy of an isolated system is constant.*

The first law of thermodynamics is an expression of the principle of conservation of energy. It states that energy can be transformed (changed from one form to another), but cannot be created or destroyed.

The first law is usually formulated by saying that the change in the internal energy of a closed thermodynamic system is equal to the difference between the heat supplied to the system and the amount of work done by the system on its surroundings. It is important to note that internal energy is a state of the system whereas heat and work modify the state of the system. In other words, an change of internal energy of a system may be achieved by any combination of heat and work added or removed from the system as long as those total to the change of internal energy. The manner by which a system achieves its internal energy is path independent.

- Second law of thermodynamics: *Heat cannot spontaneously flow from a colder location to a hotter location.*

The second law of thermodynamics is an expression of the universal principle of decay observable in nature. The second law is an observation of the fact that over time, differences in temperature, pressure, and chemical potential tend to even out in a physical system that is isolated from the outside world. Entropy is a measure of how much this process has progressed. The entropy of an isolated system which is not in equilibrium will tend to increase over time, approaching a maximum value at equilibrium.

In classical thermodynamics, the second law is a basic postulate applicable to any system involving heat energy transfer; in statistical thermodynamics, the second law is a consequence of the assumed randomness of molecular chaos. There are many versions of the second law, but they all have the same effect, which is to explain the phenomenon of irreversibility in nature.

- Third law of thermodynamics: *As a system approaches absolute zero, all processes cease and the entropy of the system approaches a minimum value.*

The third law of thermodynamics is a statistical law of nature regarding entropy and the impossibility of reaching absolute zero of temperature. This law provides an absolute reference point for the determination of entropy. The entropy determined relative to this point is the absolute entropy. Alternate definitions are, "the entropy of all systems and of all states of a system is smallest at absolute zero," or equivalently "it is impossible to reach the absolute zero of temperature by any finite number of processes".

Absolute zero, at which all activity would stop if it were possible to happen, is −273.15°C (degrees Celsius), or −459.67°F (degrees Fahrenheit) or 0 K (kelvin).

System Models

An important concept in thermodynamics is the thermodynamic system, which is a precisely defined region of the universe under study. Everything in the universe except the system is called the *surroundings*. A system is separated from the remainder of the universe by a *boundary* which may be a physical boundary or notional, but which by convention defines a finite volume. Exchanges of work, heat, or matter between the system and the surroundings take place across this boundary.

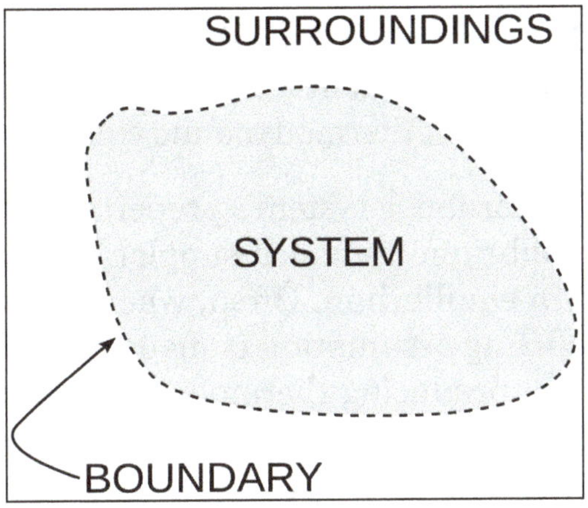

A diagram of a generic thermodynamic system.

In practice, the boundary of a system is simply an imaginary dotted line drawn around a volume within which is going to be a change in the internal energy of that volume. Anything that passes across the boundary that effects a change in the internal energy of the system needs to be accounted for in the energy balance equation. The volume can be the region surrounding a single atom resonating energy, such as Max Planck defined in 1900; it can be a body of steam or air in a steam engine, such as Sadi Carnot defined in 1824; it can be the body of a tropical cyclone, such as Kerry Emanuel theorized in 1986 in the field of atmospheric thermodynamics; it could also be just one nuclide (i.e. a system of quarks) as hypothesized in quantum thermodynamics, or the event horizon of a black hole.

Boundaries are of four types: fixed, movable, real, and imaginary. For example, in an engine, a fixed boundary means the piston is locked at its position, within which a constant volume process might occur. If the piston is allowed to move that boundary is movable while the cylinder and cylinder head boundaries are fixed. For closed systems, boundaries are real while for open systems boundaries are often imaginary. In the case of a jet engine, a fixed imaginary boundary might be assumed at the intake of the engine, fixed boundaries along the surface of the case and a second fixed imaginary boundary across the exhaust nozzle.

Generally, thermodynamics distinguishes three classes of systems, defined in terms of what is allowed to cross their boundaries:

Interactions of Thermodynamic Systems			
Type of system	**Mass flow**	**Work**	**Heat**
Open	✓	✓	✓
Closed	✗	✓	✓

Thermally isolated	✗	✓	✗
Mechanically isolated	✗	✗	✓
Isolated	✗	✗	✗

As time passes in an isolated system, internal differences of pressures, densities, and temperatures tend to even out. A system in which all equalizing processes have gone to completion is said to be in a state of thermodynamic equilibrium.

Once in thermodynamic equilibrium, a system's properties are, by definition, unchanging in time. Systems in equilibrium are much simpler and easier to understand than are systems which are not in equilibrium. Often, when analysing a dynamic thermodynamic process, the simplifying assumption is made that each intermediate state in the process is at equilibrium, producing thermodynamic processes which develop so slowly as to allow each intermediate step to be an equilibrium state and are said to be reversible processes.

States and Processes

When a system is at equilibrium under a given set of conditions, it is said to be in a definite thermodynamic state. The state of the system can be described by a number of state quantities that do not depend on the process by which the system arrived at its state. They are called intensive variables or extensive variables according to how they change when the size of the system changes. The properties of the system can be described by an equation of state which specifies the relationship between these variables. State may be thought of as the instantaneous quantitative description of a system with a set number of variables held constant.

A thermodynamic process may be defined as the energetic evolution of a thermodynamic system proceeding from an initial state to a final state. It can be described by process quantities. Typically, each thermodynamic process is distinguished from other processes in energetic character according to what parameters, such as temperature, pressure, or volume, etc., are held fixed. Furthermore, it is useful to group these processes into pairs, in which each variable held constant is one member of a conjugate pair.

Several commonly studied thermodynamic processes are:

- Adiabatic process: occurs without loss or gain of energy by heat.
- Isenthalpic process: occurs at a constant enthalpy.
- Isentropic process: a reversible adiabatic process, occurs at a constant entropy.
- Isobaric process: occurs at constant pressure.
- Isochoric process: occurs at constant volume (also called isometric/isovolumetric).

- Isothermal process: occurs at a constant temperature.
- Steady state process: occurs without a change in the internal energy.

Instrumentation

There are two types of thermodynamic instruments, the meter and the reservoir. A thermodynamic meter is any device which measures any parameter of a thermodynamic system. In some cases, the thermodynamic parameter is actually defined in terms of an idealized measuring instrument. For example, the zeroth law states that if two bodies are in thermal equilibrium with a third body, they are also in thermal equilibrium with each other. This principle, as noted by James Maxwell in 1872, asserts that it is possible to measure temperature. An idealized thermometer is a sample of an ideal gas at constant pressure. From the ideal gas law $pV=nRT$, the volume of such a sample can be used as an indicator of temperature; in this manner it defines temperature. Although pressure is defined mechanically, a pressure-measuring device, called a barometer may also be constructed from a sample of an ideal gas held at a constant temperature. A calorimeter is a device which is used to measure and define the internal energy of a system.

A thermodynamic reservoir is a system which is so large that its state parameters are not appreciably altered when it is brought into contact with the system of interest. When the reservoir is brought into contact with the system, the system is brought into equilibrium with the reservoir. For example, a pressure reservoir is a system at a particular pressure, which imposes that pressure upon the system to which it is mechanically connected. The Earth's atmosphere is often used as a pressure reservoir. If ocean water is used to cool a power plant, the ocean is often a temperature reservoir in the analysis of the power plant cycle.

Conjugate Variables

The central concept of thermodynamics is that of energy, the ability to do work. By the First Law, the total energy of a system and its surroundings is conserved. Energy may be transferred into a system by heating, compression, or addition of matter, and extracted from a system by cooling, expansion, or extraction of matter. In mechanics, for example, energy transfer equals the product of the force applied to a body and the resulting displacement.

Conjugate variables are pairs of thermodynamic concepts, with the first being akin to a "force" applied to some thermodynamic system, the second being akin to the resulting "displacement," and the product of the two equalling the amount of energy transferred. The common conjugate variables are:

- Pressure-volume (the mechanical parameters).
- Temperature-entropy (thermal parameters).
- Chemical potential-particle number (material parameters).

Potentials

Thermodynamic potentials are different quantitative measures of the stored energy in a system. Potentials are used to measure energy changes in systems as they evolve from an initial state to a final state. The potential used depends on the constraints of the system, such as constant temperature or pressure. For example, the Helmholtz and Gibbs energies are the energies available in a system to do useful work when the temperature and volume or the pressure and temperature are fixed, respectively.

The five most well known potentials are:

Name	Symbol	Formula	Natural variables
Internal energy	U	$\int (TdS - pdV + \sum_i \mu_i dN_i)$	$S, V, \{N_i\}$
Helmholtz free energy	F	$U - TS$	$T, V, \{N_i\}$
Enthalpy	H	$U + pV$	$S, p, \{N_i\}$
Gibbs free energy	G	$U + pV - TS$	$T, p, \{N_i\}$
Landau Potential (Grand potential)	Ω, Φ_G	$U - TS - \sum_i \mu_i N_i$	$T, V, \{\mu_i\}$

where T is the temperature, S the entropy, p the pressure, V the volume, μ the chemical potential, N the number of particles in the system, and i is the count of particles types in the system.

Thermodynamic potentials can be derived from the energy balance equation applied to a thermodynamic system. Other thermodynamic potentials can also be obtained through Legendre transformation.

Applied Fields

- Atmospheric thermodynamics.
- Biological thermodynamics.
- Black hole thermodynamics.
- Chemical thermodynamics.
- Classical thermodynamics.
- Equilibrium thermodynamics.

- Industrial ecology (re: Exergy).
- Maximum entropy thermodynamics.
- Non-equilibrium thermodynamics.
- Philosophy of thermal and statistical physics.
- Psychrometrics.
- Quantum thermodynamics.
- Statistical thermodynamics.
- Thermoeconomics.

Enthalpy and Internal Energy

Specific internal energy of the mixture, $u_{mix} = \sum_i X_i u_i$

Mass specific internal energy of the mixture, $\hat{u}_{mix} = \sum_i Y_i \hat{u}_i$

Specific enthalpy ofthe mixture, $h_{mix} = \sum_i X_i h_i$

Mass specific enthalpy of the mixture, $\hat{h}_{mix} = \sum_i Y_i \hat{h}_i$

Enthalpy of a species, $h^0_{i,T}(T) = \underbrace{h^0_{f,298.15}}_{\text{Heat of formation (due to bond energy)}} + \int_{298.15}^{T} \underbrace{C_{p,i}}_{\text{Sensible enthalpy (associated with temperature)}} dT$

Internal Energy of a species, $u^0_{i,T}(T) = u^0_{f,298.15} + \int_{298.15}^{T} C_{v,i} dT$

Effect of Temperature on Heat Capacity

Specific heats, C_p and C_v are functions of temperaturc for both ideal and real gases,

Variation in sp. heat with temperature is caused by

- Vibrational energy
- Rotational energy

Sp. heat of mono atomic gases does not vary with temperature, why?

Their internal energy is contributed by translational energy.

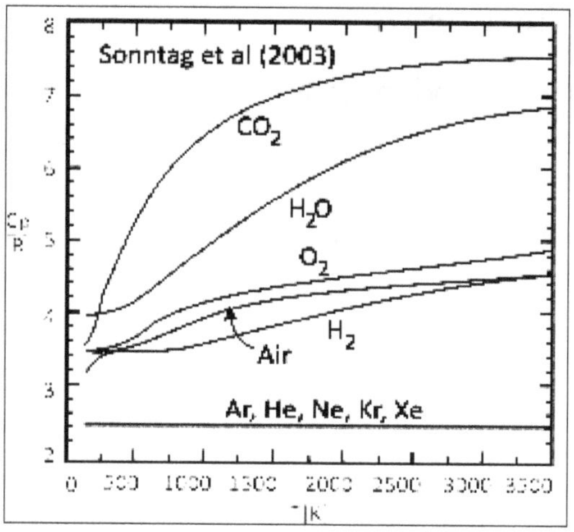

Laws of Thermodynamics

The four laws of thermodynamics define fundamental physical quantities (temperature, energy, and entropy) that characterize thermodynamic systems at thermal equilibrium. The laws describe how these quantities behave under various circumstances, and forbid certain phenomena (such as perpetual motion).

The four laws of thermodynamics are:

- Zeroth law of thermodynamics: If two systems are in thermal equilibrium with a third system, they are in thermal equilibrium with each other. This law helps define the notion of temperature.
- First law of thermodynamics: When energy passes, as work, as heat, or with matter, into or out from a system, the system's internal energy changes in accord with the law of conservation of energy. Equivalently, perpetual motion machines of the first kind are impossible.
- Second law of thermodynamics: In a natural thermodynamic process, the sum of the entropies of the interacting thermodynamic systems increases. Equivalently, perpetual motion machines of the second kind are impossible.
- Third law of thermodynamics: The entropy of a system approaches a constant value as the temperature approaches absolute zero. With the exception of non-crystalline solids (glasses) the entropy of a system at absolute zero is typically close to zero, and is equal to the logarithm of the product of the quantum ground states.

There have been suggestions of additional laws, but none of them achieves the generality of the four accepted laws, and they are not mentioned in standard textbooks.

The laws of thermodynamics are important fundamental laws in physics and they are applicable in other natural sciences.

Zeroth Law

The zeroth law of thermodynamics may be stated in the following form:

If two systems are both in thermal equilibrium with a third system then they are in thermal equilibrium with each other.

The law is intended to allow the existence of an empirical parameter, the temperature, as a property of a system such that systems in thermal equilibrium with each other have the same temperature. The law as stated here is compatible with the use of a particular physical body, for example a mass of gas, to match temperatures of other bodies, but does not justify regarding temperature as a quantity that can be measured on a scale of real numbers.

Though this version of the law is one of the more commonly stated, it is only one of a diversity of statements that are labeled as "the zeroth law" by competent writers. Some statements go further so as to supply the important physical fact that temperature is one-dimensional, that one can conceptually arrange bodies in real number sequence from colder to hotter. Perhaps there exists no unique "best possible statement" of the "zeroth law", because there is in the literature a range of formulations of the principles of thermodynamics, each of which call for their respectively appropriate versions of the law.

Although these concepts of temperature and of thermal equilibrium are fundamental to thermodynamics and were clearly stated in the nineteenth century, the desire to explicitly number the above law was not widely felt until Fowler and Guggenheim did so in the 1930s, long after the first, second, and third law were already widely understood and recognized. Hence it was numbered the zeroth law. The importance of the law as a foundation to the earlier laws is that it allows the definition of temperature in a non-circular way without reference to entropy, its conjugate variable. Such a temperature definition is said to be 'empirical'.

First Law

The first law of thermodynamics may be stated in several ways :

> The increase in internal energy of a closed system is equal to total of the energy added to the system. In particular, if the energy entering the system is supplied as heat and energy leaves the system as work, the heat is accounted as positive and the work is accounted as negative.
>
> $$\Delta U_{system} = Q - W$$
>
> In the case of a thermodynamic cycle of a closed system, which returns to its original state, the heat Q_{in} supplied to the system in one stage of the cycle, minus

the heat Q_{out} removed from it in another stage of the cycle, plus the work added to the system W_{in} equals the work that leaves the system W_{out}.

$$\Delta U_{system(full\ cycle)} = 0$$

hence, for a full cycle,

$$Q = Q_{in} - Q_{out} + W_{in} - W_{out} = W_{net}$$

For the particular case of a thermally isolated system (adiabatically isolated), the change of the internal energy of an adiabatically isolated system can only be the result of the work added to the system, because the adiabatic assumption is: $Q = 0$.

$$\Delta U_{system} = U_{final} - U_{initial} = W_{in} - W_{out}$$

More specifically, the First Law encompasses several principles:

- The law of conservation of energy.

  This states that energy can be neither created nor destroyed. However, energy can change forms, and energy can flow from one place to another. A particular consequence of the law of conservation of energy is that the total energy of an isolated system does not change.

- The concept of internal energy and its relationship to temperature.

  If a system has a definite temperature, then its total energy has three distinguishable components. If the system is in motion as a whole, it has kinetic energy. If the system as a whole is in an externally imposed force field (e.g. gravity), it has potential energy relative to some reference point in space. Finally, it has internal energy, which is a fundamental quantity of thermodynamics. The establishment of the concept of internal energy distinguishes the first law of thermodynamics from the more general law of conservation of energy.

$$E_{total} = \mathrm{KE}_{system} + \mathrm{PE}_{system} + U_{system}$$

The internal energy of a substance can be explained as the sum of the diverse kinetic energies of the erratic microscopic motions of its constituent atoms, and of the potential energy of interactions between them. Those microscopic energy terms are collectively called the substance's internal energy (U), and are accounted for by macroscopic thermodynamic property. The total of the kinetic energies of microscopic motions of the constituent atoms increases as the system's temperature increases; this assumes no other interactions at the microscopic level of the system such as chemical reactions, potential energy of constituent atoms with respect to each other.

- Work is a process of transferring energy to or from a system in ways that can be described by macroscopic mechanical forces exerted by factors in the surroundings, outside the system. Examples are an externally driven shaft agitating a stirrer within the system, or an externally imposed electric field that polarizes the material of the system, or a piston that compresses the system. Unless otherwise stated, it is customary to treat work as occurring without its dissipation to the surroundings. Practically speaking, in all natural process, some of the work is dissipated by internal friction or viscosity. The work done by the system can come from its overall kinetic energy, from its overall potential energy, or from its internal energy.

  For example, when a machine (not a part of the system) lifts a system upwards, some energy is transferred from the machine to the system. The system's energy increases as work is done on the system and in this particular case, the energy increase of the system is manifested as an increase in the system's gravitational potential energy. Work added to the system increases the Potential Energy of the system:

$$W = \Delta \mathrm{PE}_{system}$$

  Or in general, the energy added to the system in the form of work can be partitioned to kinetic, potential or internal energy forms:

$$W = \Delta \mathrm{KE}_{system} + \Delta \mathrm{PE}_{system} + \Delta U_{system}$$

- When matter is transferred into a system, that masses' associated internal energy and potential energy are transferred with it.

$$\left(u \Delta M\right)_{in} = \Delta U_{system}$$

  where u denotes the internal energy per unit mass of the transferred matter, as measured while in the surroundings; and ΔM denotes the amount of transferred mass.

- The flow of heat is a form of energy transfer.

  Heating is a natural process of moving energy to or from a system other than by work or the transfer of matter. Direct passage of heat is only from a hotter to a colder system.

  If the system has rigid walls that are impermeable to matter, and consequently energy cannot be transferred as work into or out from the system, and no external long-range force field affects it that could change its internal energy, then the internal energy can only be changed by the transfer of energy as heat:

$$\Delta U_{system} = Q$$

where Q denotes the amount of energy transferred into the system as heat.

Combining these principles leads to one traditional statement of the first law of thermodynamics: it is not possible to construct a machine which will perpetually output work without an equal amount of energy input to that machine. Or more briefly, a perpetual motion machine of the first kind is impossible.

Second Law

The second law of thermodynamics indicates the irreversibility of natural processes, and, in many cases, the tendency of natural processes to lead towards spatial homogeneity of matter and energy, and especially of temperature. It can be formulated in a variety of interesting and important ways.

It implies the existence of a quantity called the entropy of a thermodynamic system. In terms of this quantity it implies that:

When two initially isolated systems in separate but nearby regions of space, each in thermodynamic equilibrium with itself but not necessarily with each other, are then allowed to interact, they will eventually reach a mutual thermodynamic equilibrium. The sum of the entropies of the initially isolated systems is less than or equal to the total entropy of the final combination. Equality occurs just when the two original systems have all their respective intensive variables (temperature, pressure) equal; then the final system also has the same values.

This statement of the second law is founded on the assumption, that in classical thermodynamics, the entropy of a system is defined only when it has reached internal thermodynamic equilibrium (thermodynamic equilibrium with itself).

The second law is applicable to a wide variety of processes, reversible and irreversible. All natural processes are irreversible. Reversible processes are a useful and convenient theoretical fiction, but do not occur in nature.

A prime example of irreversibility is in the transfer of heat by conduction or radiation. It was known long before the discovery of the notion of entropy that when two bodies initially of different temperatures come into thermal connection, then heat always flows from the hotter body to the colder one.

The second law tells also about kinds of irreversibility other than heat transfer, for example those of friction and viscosity, and those of chemical reactions. The notion of entropy is needed to provide that wider scope of the law.

According to the second law of thermodynamics, in a theoretical and fictive reversible heat transfer, an element of heat transferred, δQ, is the product of the temperature (T), both of the system and of the sources or destination of the heat, with the increment (dS) of the system's conjugate variable, its entropy (S)

$$\delta Q = T\,dS$$

Entropy may also be viewed as a physical measure of the lack of physical information about the microscopic details of the motion and configuration of a system, when only the macroscopic states are known. This lack of information is often described as *disorder* on a microscopic or molecular scale. The law asserts that for two given macroscopically specified states of a system, there is a quantity called the difference of information entropy between them. This information entropy difference defines how much additional microscopic physical information is needed to specify one of the macroscopically specified states, given the macroscopic specification of the other - often a conveniently chosen reference state which may be presupposed to exist rather than explicitly stated. A final condition of a natural process always contains microscopically specifiable effects which are not fully and exactly predictable from the macroscopic specification of the initial condition of the process. This is why entropy increases in natural processes - the increase tells how much extra microscopic information is needed to distinguish the final macroscopically specified state from the initial macroscopically specified state.

Third Law

The third law of thermodynamics is sometimes stated as follows:

> *The entropy of a perfect crystal of any pure substance approaches zero as the temperature approaches absolute zero.*

At zero temperature the system must be in a state with the minimum thermal energy. This statement holds true if the perfect crystal has only one state with minimum energy. Entropy is related to the number of possible microstates according to:

$$S = k_B \ln \Omega$$

Where S is the entropy of the system, k_B Boltzmann's constant, and Ω the number of microstates (e.g. possible configurations of atoms). At absolute zero there is only 1 microstate possible (Ω=1 as all the atoms are identical for a pure substance and as a result all orders are identical as there is only one combination) and ln(1) = 0.

A more general form of the third law that applies to a system such as a glass that may have more than one minimum microscopically distinct energy state, or may have a microscopically distinct state that is "frozen in" though not a strictly minimum energy state and not strictly speaking a state of thermodynamic equilibrium, at absolute zero temperature:

> *The entropy of a system approaches a constant value as the temperature approaches zero.*

The constant value (not necessarily zero) is called the residual entropy of the system.

History

Circa 1797, Count Rumford (born Benjamin Thompson) showed that endless mechanical action can generate indefinitely large amounts of heat from a fixed amount of working substance thus challenging the caloric theory of heat, which held that there would be a finite amount of caloric heat/energy in a fixed amount of working substance. The first established thermodynamic principle, which eventually became the second law of thermodynamics, was formulated by Sadi Carnot in 1824. By 1860, as formalized in the works of those such as Rudolf Clausius and William Thomson, two established principles of thermodynamics had evolved, the first principle and the second principle, later restated as thermodynamic laws. By 1873, for example, thermodynamicist Josiah Willard Gibbs, in his memoir *Graphical Methods in the Thermodynamics of Fluids*, clearly stated the first two absolute laws of thermodynamics. Some textbooks throughout the 20th century have numbered the laws differently. In some fields removed from chemistry, the second law was considered to deal with the efficiency of heat engines only, whereas what was called the third law dealt with entropy increases. Directly defining zero points for entropy calculations was not considered to be a law. Gradually, this separation was combined into the second law and the modern third law was widely adopted.

Stoichiometry

Stoichiometry is the calculation of relative quantities of reactants and products in chemical reactions.

Stoichiometry is founded on the law of conservation of mass where the total mass of the reactants equals the total mass of the products leading to the insight that the relations among quantities of reactants and products typically form a ratio of positive integers. This means that if the amounts of the separate reactants are known, then the amount of the product can be calculated. Conversely, if one reactant has a known quantity and the quantity of product can be empirically determined, then the amount of the other reactants can also be calculated.

This is illustrated in the image here, where the balanced equation is:

$$CH_4 + 2\,O_2 \rightarrow CO_2 + 2\,H_2O.$$

Here, one molecule of methane reacts with two molecules of oxygen gas to yield one molecule of carbon dioxide and two molecules of water. Stoichiometry measures these quantitative relationships, and is used to determine the amount of products/ reactants that are produced/needed in a given reaction. Describing the quantitative relationships among substances as they participate in chemical reactions is known as *reaction stoichiometry*. In the example above, reaction stoichiometry measures the relationship between the methane and oxygen as they react to form carbon dioxide and water.

Because of the well known relationship of moles to atomic weights, the ratios that are arrived at by stoichiometry can be used to determine quantities by weight in a reaction described by a balanced equation. This is called *composition stoichiometry*.

Gas stoichiometry deals with reactions involving gases, where the gases are at a known temperature, pressure, and volume and can be assumed to be ideal gases. For gases, the volume ratio is ideally the same by the ideal gas law, but the mass ratio of a single reaction has to be calculated from the molecular masses of the reactants and products. In practice, due to the existence of isotopes, molar masses are used instead when calculating the mass ratio.

Etymology

The term *stoichiometry* was first used by Jeremias Benjamin Richter in 1792 when the first volume of Richter's *Stoichiometry or the Art of Measuring the Chemical Elements* was published. The term is derived from the Greek words *stoicheion* "element" and *metron* "measure". In patristic Greek, the word *Stoichiometria* was used by Nicephorus to refer to the number of line counts of the canonical New Testament and some of the Apocrypha.

Definition

A stoichiometric amount or stoichiometric ratio of a reagent is the optimum amount or ratio where, assuming that the reaction proceeds to completion:

1. All of the reagent is consumed.
2. There is no deficiency of the reagent.
3. There is no cxcess of the reagent.

Stoichiometry rests upon the very basic laws that help to understand it better, i.e., law of conservation of mass, the law of definite proportions (i.e., the law of constant composition), the law of multiple proportions and the law of reciprocal proportions. In general, chemical reactions combine in definite ratios of chemicals. Since chemical reactions can neither create nor destroy matter, nor transmute one element into another, the amount of each element must be the same throughout the overall reaction. For example, the number of atoms of a given element X on the reactant side must equal the number of atoms of that element on the product side, whether or not all of those atoms are actually involved in a reaction.

Chemical reactions, as macroscopic unit operations, consist of simply a very large number of elementary reactions, where a single molecule reacts with another molecule. As the reacting molecules (or moieties) consist of a definite set of atoms in an integer ratio, the ratio between reactants in a complete reaction is also in integer ratio. A reaction may consume more than one molecule, and the stoichiometric number counts this number, defined as positive for products (added) and negative for reactants (removed).

Different elements have a different atomic mass, and as collections of single atoms, molecules have a definite molar mass, measured with the unit mole (6.02×10^{23} individual molecules, Avogadro's constant). By definition, carbon-12 has a molar mass of 12 g/mol. Thus, to calculate the stoichiometry by mass, the number of molecules required for each reactant is expressed in moles and multiplied by the molar mass of each to give the mass of each reactant per mole of reaction. The mass ratios can be calculated by dividing each by the total in the whole reaction.

Elements in their natural state are mixtures of isotopes of differing mass, thus atomic masses and thus molar masses are not exactly integers. For instance, instead of an exact 14:3 proportion, 17.04 kg of ammonia consists of 14.01 kg of nitrogen and 3×1.01 kg of hydrogen, because natural nitrogen includes a small amount of nitrogen-15, and natural hydrogen includes hydrogen-2 (deuterium).

A stoichiometric reactant is a reactant that is consumed in a reaction, as opposed to a catalytic reactant, which is not consumed in the overall reaction because it reacts in one step and is regenerated in another step.

Converting Grams to Moles

Stoichiometry is not only used to balance chemical equations but also used in conversions, i.e., converting from grams to moles using molar mass as the conversion factor, or from grams to milliliters using density. For example, to find the amount of NaCl (sodium chloride) in 2.00 g, one would do the following:

$$\frac{2.00 \text{ g NaCl}}{58.44 \text{ g NaCl mol}^{-1}} = 0.034 \text{ mol}$$

In the above example, when written out in fraction form, the units of grams form a multiplicative identity, which is equivalent to one (g/g = 1), with the resulting amount in moles (the unit that was needed), as shown in the following equation,

$$\left(\frac{2.00 \text{ g NaCl}}{1}\right)\left(\frac{1 \text{ mol NaCl}}{58.44 \text{ g NaCl}}\right) = 0.034 \text{ mol}$$

Molar Proportion

Stoichiometry is often used to balance chemical equations (reaction stoichiometry). For example, the two diatomic gases, hydrogen and oxygen, can combine to form a liquid, water, in an exothermic reaction, as described by the following equation:

$$2\ H_2 + O_2 \rightarrow 2\ H_2O$$

Reaction stoichiometry describes the 2:1:2 ratio of hydrogen, oxygen, and water molecules in the above equation.

The molar ratio allows for conversion between moles of one substance and moles of another. For example, in the reaction

$$2\ CH_3OH + 3\ O_2 \rightarrow 2\ CO_2 + 4\ H_2O$$

the amount of water that will be produced by the combustion of 0.27 moles of CH_3OH is obtained using the molar ratio between CH_3OH and H_2O of 2 to 4.

$$\left(\frac{0.27\ \text{mol}\ CH_3OH}{1}\right)\left(\frac{4\ \text{mol}\ H_2O}{2\ \text{mol}\ CH_3OH}\right) = 0.54\ \text{mol}\ H_2O$$

The term stoichiometry is also often used for the molar proportions of elements in stoichiometric compounds (composition stoichiometry). For example, the stoichiometry of hydrogen and oxygen in H_2O is 2:1. In stoichiometric compounds, the molar proportions are whole numbers.

Determining Amount of Product

Stoichiometry can also be used to find the quantity of a product yielded by a reaction. If a piece of solid copper (Cu) were added to an aqueous solution of silver nitrate ($AgNO_3$), the silver (Ag) would be replaced in a single displacement reaction forming aqueous copper(II) nitrate ($Cu(NO_3)_2$) and solid silver. How much silver is produced if 16.00 grams of Cu is added to the solution of excess silver nitrate?

The following steps would be used:

1. Write and balance the equation.
2. Mass to moles: Convert grams of Cu to moles of Cu.
3. Mole ratio: Convert moles of Cu to moles of Ag produced.
4. Mole to mass: Convert moles of Ag to grams of Ag produced.

The complete balanced equation would be:

$$Cu + 2\ AgNO_3 \rightarrow Cu(NO_3)_2 + 2\ Ag$$

For the mass to mole step, the mass of copper (16.00 g) would be converted to moles of copper by dividing the mass of copper by its molecular mass: 63.55 g/mol.

$$\left(\frac{16.00\ \text{g Cu}}{1}\right)\left(\frac{1\ \text{mol Cu}}{63.55\ \text{g Cu}}\right) = 0.2518\ \text{mol Cu}$$

Now that the amount of Cu in moles (0.2518) is found, we can set up the mole ratio. This is found by looking at the coefficients in the balanced equation: Cu and Ag are in a 1:2 ratio.

$$\left(\frac{0.2518\ \text{mol Cu}}{1}\right)\left(\frac{2\ \text{mol Ag}}{1\ \text{mol Cu}}\right) = 0.5036\ \text{mol Ag}$$

Now that the moles of Ag produced is known to be 0.5036 mol, we convert this amount to grams of Ag produced to come to the final answer:

$$\left(\frac{0.5036\ \text{mol Ag}}{1}\right)\left(\frac{107.87\ \text{g Ag}}{1\ \text{mol Ag}}\right) = 54.32\ \text{g Ag}$$

This set of calculations can be further condensed into a single step:

$$m_{\text{Ag}} = \left(\frac{16.00\ \text{g Cu}}{1}\right)\left(\frac{1\ \text{mol Cu}}{63.55\ \text{g Cu}}\right)\left(\frac{2\ \text{mol Ag}}{1\ \text{mol Cu}}\right)\left(\frac{107.87\ \text{g Ag}}{1\ \text{mol Ag}}\right) = 54.32\ \text{g}$$

Further Examples

For propane (C_3H_8) reacting with oxygen gas (O_2), the balanced chemical equation is:

$$C_3H_8 + 5\,O_2 \rightarrow 3\,CO_2 + 4\,H_2O$$

The mass of water formed if 120 g of propane (C_3H_8) is burned in excess oxygen is then,

$$m\text{H}_2\text{O} = \left(\frac{120.\ \text{g C}_3\text{H}_8}{1}\right)\left(\frac{1\ \text{mol C}_3\text{H}_8}{44.09\ \text{g C}_3\text{H}_8}\right)\left(\frac{4\ \text{mol H}_2\text{O}}{1\ \text{mol C}_3\text{H}_8}\right)\left(\frac{18.02\ \text{g H}_2\text{O}}{1\ \text{mol H}_2\text{O}}\right) = 196\ \text{g}$$

Stoichiometric Ratio

Stoichiometry is also used to find the right amount of one reactant to "completely" react with the other reactant in a chemical reaction – that is, the stoichiometric amounts that would result in no leftover reactants when the reaction takes place. An example is shown below using the thermite reaction,

$$Fe_2O_3 + 2\,Al \rightarrow Al_2O_3 + 2\,Fe$$

This equation shows that 1 mole of iron(III) oxide and 2 moles of aluminum will produce 1 mole of aluminium oxide and 2 moles of iron. So, to completely react with 85.0 g of iron(III) oxide (0.532 mol), 28.7 g (1.06 mol) of aluminium are needed.

$$m_{Al} = \left(\frac{85.0 \text{ g } Fe_2O_3}{1}\right)\left(\frac{1 \text{ mol } Fe_2O_3}{159.7 \text{ g } Fe_2O_3}\right)\left(\frac{2 \text{ mol Al}}{1 \text{ mol } Fe_2O_3}\right)\left(\frac{26.98 \text{ g Al}}{1 \text{ mol Al}}\right) = 28.7 \text{ g}$$

Limiting Reagent and Percent Yield

The limiting reagent is the reagent that limits the amount of product that can be formed and is completely consumed when the reaction is complete. An excess reactant is a reactant that is left over once the reaction has stopped due to the limiting reactant being exhausted.

Consider the equation of roasting lead(II) sulfide (PbS) in oxygen (O_2) to produce lead(II) oxide (PbO) and sulfur dioxide (SO_2):

$$2\ PbS + 3\ O_2 \rightarrow 2\ PbO + 2\ SO_2$$

To determine the theoretical yield of lead(II) oxide if 200.0 g of lead(II) sulfide and 200.0 g of oxygen are heated in an open container:

$$m_{PbO} = \left(\frac{200.0 \text{ g PbS}}{1}\right)\left(\frac{1 \text{ mol PbS}}{239.27 \text{ g PbS}}\right)\left(\frac{2 \text{ mol PbO}}{2 \text{ mol PbS}}\right)\left(\frac{223.2 \text{ g PbO}}{1 \text{ mol PbO}}\right) = 186.6 \text{ g}$$

$$m_{PbO} = \left(\frac{200.0 \text{ g } O_2}{1}\right)\left(\frac{1 \text{ mol } O_2}{32.00 \text{ g } O_2}\right)\left(\frac{2 \text{ mol PbO}}{3 \text{ mol } O_2}\right)\left(\frac{223.2 \text{ g PbO}}{1 \text{ mol PbO}}\right) = 930.0 \text{ g}$$

Because a lesser amount of PbO is produced for the 200.0 g of PbS, it is clear that PbS is the limiting reagent.

In reality, the actual yield is not the same as the stoichiometrically-calculated theoretical yield. Percent yield, then, is expressed in the following equation:

$$\text{percent yield} = \frac{\text{actual yield}}{\text{theoretical yield}}$$

If 170.0 g of lead(II) oxide is obtained, then the percent yield would be calculated as follows:

$$\text{percent yield} = \frac{170.0 \text{ g PbO}}{186.6 \text{ g PbO}} = 91.12\%$$

Example

Consider the following reaction, in which iron(III) chloride reacts with hydrogen sulfide to produce iron(III) sulfide and hydrogen chloride:

$$2\ FeCl_3 + 3\ H_2S \rightarrow Fe_2S_3 + 6\ HCl$$

Suppose 90.0 g of $FeCl_3$ reacts with 52.0 g of H_2S. To find the limiting reagent and the mass of HCl produced by the reaction, we could set up the following equations:

$$m_{\text{HCl}} = \left(\frac{90.0\text{ g FeCl}_3}{1}\right)\left(\frac{1\text{ mol FeCl}_3}{162\text{ g FeCl}_3}\right)\left(\frac{6\text{ mol HCl}}{2\text{ mol FeCl}_3}\right)\left(\frac{36.5\text{ g HCl}}{1\text{ mol HCl}}\right) = 60.8\text{ g}$$

$$m_{\text{HCl}} = \left(\frac{52.0\text{ g H}_2\text{S}}{1}\right)\left(\frac{1\text{ mol H}_2\text{S}}{34.1\text{ g H}_2\text{S}}\right)\left(\frac{6\text{ mol HCl}}{3\text{ mol H}_2\text{S}}\right)\left(\frac{36.5\text{ g HCl}}{1\text{ mol HCl}}\right) = 111\text{ g}$$

Thus, the limiting reagent is $FeCl_3$ and the amount of HCl produced is 60.8 g.

To find what mass of excess reagent (H_2S) remains after the reaction, we would set up the calculation to find out how much H_2S reacts completely with the 90.0 g $FeCl_3$:

$$m_{\text{H}_2\text{S}} = \left(\frac{90.0\text{ g FeCl}_3}{1}\right)\left(\frac{1\text{ mol FeCl}_3}{162\text{ g FeCl}_3}\right)\left(\frac{3\text{ mol H}_2\text{S}}{2\text{ mol FeCl}_3}\right)\left(\frac{34.1\text{ g H}_2\text{S}}{1\text{ mol H}_2\text{S}}\right) = 28.4\text{ g reacted}$$

By subtracting this amount from the original amount of H_2S, we can come to the answer:

$$52.0\text{ g H}_2\text{S} - 28.4\text{ g H}_2\text{S} = 23.6\text{ g H}_2\text{S excess}$$

Different Stoichiometries in Competing Reactions

Often, more than one reaction is possible given the same starting materials. The reactions may differ in their stoichiometry. For example, the methylation of benzene (C_6H_6), through a Friedel–Crafts reaction using $AlCl_3$ as a catalyst, may produce singly methylated ($C_6H_5CH_3$), doubly methylated ($C_6H_4(CH_3)_2$), or still more highly methylated ($C_6H_{6-n}(CH_3)_n$) products, as shown in the following example,

$$C_6H_6 + CH_3Cl \rightarrow C_6H_5CH_3 + HCl$$

$$C_6H_6 + 2\ CH_3Cl \rightarrow C_6H_4(CH_3)_2 + 2\ HCl$$

$$C_6H_6 + n\ CH_3Cl \rightarrow C_6H_{6-n}(CH_3)_n + n\ HCl$$

In this example, which reaction takes place is controlled in part by the relative concentrations of the reactants.

Stoichiometric Coefficient

In lay terms, the *stoichiometric coefficient* (or *stoichiometric number* in the IUPAC nomenclature) of any given component is the number of molecules that participate in the reaction as written.

For example, in the reaction $CH_4 + 2\ O_2 \rightarrow CO_2 + 2\ H_2O$, the stoichiometric coefficient of CH_4 is −1, the stoichiometric coefficient of O_2 is −2, for CO_2 it would be +1 and for H_2O it is +2.

In more technically precise terms, the stoichiometric coefficient in a chemical reaction system of the *i*th component is defined as,

$$\nu_i = \frac{\Delta N_i}{\Delta \xi}$$

or

$$\Delta N_i = \nu_i \Delta \xi$$

where N_i is the number of molecules of i, and ξ is the progress variable or extent of reaction.

The extent of reaction ξ can be regarded as [the amount of] a real (or hypothetical) product, one molecule of which produced each time the reaction event occurs. It is the extensive quantity describing the progress of a chemical reaction equal to the number of chemical transformations, as indicated by the reaction equation on a molecular scale, divided by the Avogadro constant (in essence, it is the amount of chemical transformations). The change in the extent of reaction is given by $d\xi = dn_B/v_B$, where v_B is the stoichiometric number of any reaction entity B (reactant or product) and n_B is the corresponding amount.

The stoichiometric coefficient v_i represents the degree to which a chemical species participates in a reaction. The convention is to assign negative coefficients to *reactants* (which are consumed) and positive ones to *products*. However, any reaction may be viewed as going in the reverse direction, and all the coefficients then change sign (as does the free energy). Whether a reaction actually *will* go in the arbitrarily selected forward direction or not depends on the amounts of the substances present at any given time, which determines the kinetics and thermodynamics, i.e., whether equilibrium lies to the *right* or the *left*.

In reaction mechanisms, stoichiometric coefficients for each step are always integers, since elementary reactions always involve whole molecules. If one uses a composite representation of an overall reaction, some may be rational fractions. There are often chemical species present that do not participate in a reaction; their stoichiometric coefficients are therefore zero. Any chemical species that is regenerated, such as a catalyst, also has a stoichiometric coefficient of zero.

The simplest possible case is an isomerization,

$$A \rightarrow B$$

in which $\nu_B = 1$ since one molecule of B is produced each time the reaction occurs, while $\nu_A = -1$ since one molecule of A is necessarily consumed. In any chemical reaction, not only is the total mass conserved but also the numbers of atoms of each kind are conserved, and this imposes corresponding constraints on possible values for the stoichiometric coefficients.

There are usually multiple reactions proceeding simultaneously in any natural reaction system, including those in biology. Since any chemical component can participate in several reactions simultaneously, the stoichiometric coefficient of the *i*th component in the *k*th reaction is defined as,

$$\nu_{ik} = \frac{\partial N_i}{\partial \xi_k}$$

so that the total (differential) change in the amount of the *i*th component is,

$$dN_i = \sum_k \nu_{ik} d\xi_k .$$

Extents of reaction provide the clearest and most explicit way of representing compositional change, although they are not yet widely used.

With complex reaction systems, it is often useful to consider both the representation of a reaction system in terms of the amounts of the chemicals present $\{ N_i \}$ (state variables), and the representation in terms of the actual compositional degrees of freedom, as expressed by the extents of reaction $\{ \xi_k \}$. The transformation from a vector expressing the extents to a vector expressing the amounts uses a rectangular matrix whose elements are the stoichiometric coefficients $[\nu_{ik}]$.

The maximum and minimum for any ξ_k occur whenever the first of the reactants is depleted for the forward reaction; or the first of the "products" is depleted if the reaction as viewed as being pushed in the reverse direction. This is a purely kinematic restriction on the reaction simplex, a hyperplane in composition space, or *N*space, whose dimensionality equals the number of *linearly-independent* chemical reactions. This is necessarily less than the number of chemical components, since each reaction manifests a relation between at least two chemicals. The accessible region of the hyperplane depends on the amounts of each chemical species actually present, a contingent fact. Different such amounts can even generate different hyperplanes, all sharing the same algebraic stoichiometry.

In accord with the principles of chemical kinetics and thermodynamic equilibrium, every chemical reaction is *reversible*, at least to some degree, so that each equilibrium point must be an interior point of the simplex. As a consequence, extrema for the ξs will not occur unless an experimental system is prepared with zero initial amounts of some products.

The number of *physically*-independent reactions can be even greater than the number of chemical components, and depends on the various reaction mechanisms. For example, there may be two (or more) reaction *paths* for the isomerism above. The reaction may occur by itself, but faster and with different intermediates, in the presence of a catalyst.

The (dimensionless) "units" may be taken to be molecules or moles. Moles are most commonly used, but it is more suggestive to picture incremental chemical reactions in terms of molecules. The *N*s and ξs are reduced to molar units by dividing by Avogadro's number. While dimensional mass units may be used, the comments about integers are then no longer applicable.

Stoichiometry Matrix

In complex reactions, stoichiometries are often represented in a more compact form called the stoichiometry matrix. The stoichiometry matrix is denoted by the symbol N.

If a reaction network has *n* reactions and *m* participating molecular species then the stoichiometry matrix will have correspondingly *m* rows and *n* columns.

For example, consider the system of reactions shown below:

$$S_1 \rightarrow S_2$$

$$5\,S_3 + S_2 \rightarrow 4\,S_3 + 2\,S_2$$

$$S_3 \rightarrow S_4$$

$$S_4 \rightarrow S_5$$

This systems comprises four reactions and five different molecular species. The stoichiometry matrix for this system can be written as:

$$\mathbf{N} = \begin{bmatrix} -1 & 0 & 0 & 0 \\ 1 & 1 & 0 & 0 \\ 0 & -1 & -1 & 0 \\ 0 & 0 & 1 & -1 \\ 0 & 0 & 0 & 1 \end{bmatrix}$$

where the rows correspond to S_1, S_2, S_3, S_4 and S_5, respectively. Note that the process of converting a reaction scheme into a stoichiometry matrix can be a lossy transformation, for example, the stoichiometries in the second reaction simplify when included in the matrix. This means that it is not always possible to recover the original reaction scheme from a stoichiometry matrix.

Often the stoichiometry matrix is combined with the rate vector, v, and the species vector, S to form a compact equation describing the rates of change of the molecular species:

$$\frac{d\mathbf{S}}{dt} = \mathbf{N} \cdot \mathbf{v}.$$

Gas Stoichiometry

Gas stoichiometry is the quantitative relationship (ratio) between reactants and products in a chemical reaction with reactions that produce gases. Gas stoichiometry applies when the gases produced are assumed to be ideal, and the temperature, pressure, and volume of the gases are all known. The ideal gas law is used for these calculations. Often, but not always, the standard temperature and pressure (STP) are taken as 0°C and 1 bar and used as the conditions for gas stoichiometric calculations.

Gas stoichiometry calculations solve for the unknown volume or mass of a gaseous product or reactant. For example, if we wanted to calculate the volume of gaseous NO_2 produced from the combustion of 100 g of NH_3, by the reaction:

$$4\,NH_3(g) + 7\,O_2(g) \rightarrow 4\,NO_2(g) + 6\,H_2O(l)$$

we would carry out the following calculations:

$$100\mathrm{gNH_3} \cdot \frac{1\mathrm{molNH_3}}{17.034\mathrm{gNH_3}} = 5.871\mathrm{molNH_3}$$

There is a 1:1 molar ratio of NH_3 to NO_2 in the above balanced combustion reaction, so 5.871 mol of NO_2 will be formed. We will employ the ideal gas law to solve for the volume at 0°C (273.15 K) and 1 atmosphere using the gas law constant of R = 0.08206 L·atm·K^{-1}·mol^{-1} :

$$\begin{aligned} PV &= nRT \\ V &= \frac{nRT}{P} \\ &= \frac{5.871 \cdot 0.08206 \cdot 273.15}{1} \\ &= 131.597\mathrm{LNO_2} \end{aligned}$$

Gas stoichiometry often involves having to know the molar mass of a gas, given the density of that gas. The ideal gas law can be re-arranged to obtain a relation between the density and the molar mass of an ideal gas.

$$\rho = \frac{m}{V} \quad \text{and} \quad n = \frac{m}{M}$$

and thus:

$$\rho = \frac{MP}{RT}$$

where:

- P = absolute gas pressure.
- V = gas volume.
- n = amount (measured in moles).
- R = universal ideal gas law constant.
- T = absolute gas temperature.
- ρ = gas density at T and P.
- m = mass of gas.
- M = molar mass of gas.

Stoichiometric Air-to-fuel Ratios of Common Fuels

In the combustion reaction, oxygen reacts with the fuel, and the point where exactly all oxygen is consumed and all fuel burned is defined as the stoichiometric point. With more oxygen (overstoichiometric combustion), some of it stays unreacted. Likewise, if the combustion is incomplete due to lack of sufficient oxygen, fuel remains unreacted. (Unreacted fuel may also remain because of slow combustion or insufficient mixing of fuel and oxygen – this is not due to stoichiometry). Different hydrocarbon fuels have different contents of carbon, hydrogen and other elements, thus their stoichiometry varies.

Fuel	Ratio by mass	Ratio by volume	Percent fuel by mass
Gasoline	14.7 : 1	—	6.8%
Natural gas	17.2 : 1	9.7 : 1	5.8%
Propane (LP)	15.67 : 1	23.9 : 1	6.45%
Ethanol	9 : 1	—	11.1%
Methanol	6.47 : 1	—	15.6%
n-Butanol	11.2 : 1	—	8.2%
Hydrogen	34.3 : 1	2.39 : 1	2.9%
Diesel	14.5 : 1	—	6.8%
Methane	17.19 : 1	9.52 : 1	5.5%

Gasoline engines can run at stoichiometric air-to-fuel ratio, because gasoline is quite volatile and is mixed (sprayed or carburetted) with the air prior to ignition. Diesel engines, in contrast, run lean, with more air available than simple stoichiometry would require. Diesel fuel is less volatile and is effectively burned as it is injected, leaving less time for evaporation and mixing. Thus, it would form soot (black smoke) at stoichiometric ratio.

Conservation of Mass

Russian scientist Mikhail Lomonosov discovered the law of mass conservation in 1756 by experiments, and came to the conclusion that phlogiston theory is incorrect.

Antoine Lavoisier's discovery of the Law of Conservation of Mass led to many new findings in the 19th century. Joseph Proust's law of definite proportions and John Dalton's atomic theory branched from the discoveries of Antoine Lavoisier. Lavoisier's quantitative experiments revealed that combustion involved oxygen rather than what was previously thought to be phlogiston.

The law of conservation of mass or principle of mass conservation states that for any system closed to all transfers of matter and energy, the mass of the system must remain constant over time, as system mass cannot change quantity if it is not added or removed. Hence, the quantity of mass is "conserved" over time. The law implies that mass can neither be created nor destroyed, although it may be rearranged in space, or the entities associated with it may be changed in form, as for example when light or physical work is transformed into particles that contribute the same mass to the system as the light or work had contributed. Thus, during any chemical reaction, nuclear reaction, or radioactive decay in an isolated system, the total mass of the reactants or starting materials must be equal to the mass of the products.

The concept of mass conservation is widely used in many fields such as chemistry, mechanics, and fluid dynamics. Historically, mass conservation was discovered in chemical reactions by Antoine Lavoisier in the late 18th century, and was of crucial importance in the progress from alchemy to the modern natural science of chemistry.

The closely related concept of matter conservation was found to hold good in chemistry to such high approximation that it failed only for the high energies treated by the later refinements of relativity theory, but otherwise remains useful and sufficiently accurate for most chemical calculations, even in modern practice.

In special relativity, needed for accuracy when large energy transfers between systems is involved, the difference between thermodynamically closed and isolated systems becomes important, since conservation of mass is strictly and perfectly upheld only for so-called isolated systems, i.e. those completely isolated from all exchanges with the environment. In this circumstance, the mass–energy equivalence theorem states that mass conservation is equivalent to total energy conservation, which is the first law of thermodynamics. By contrast, for a thermodynamically closed system (i.e., one which is closed to exchanges of matter, but open to exchanges of non-material energy, such as heat and work, with the surroundings) mass is (usually) only approximately conserved. The input or output of non-material energy must change the mass of the system in relativity theory, although the change is usually small, since relatively large amounts of such energy (by comparison with ordinary experience) carry only a small amount of mass (again by ordinary standards of measurement).

In special relativity, mass is not *converted* to energy, since mass and energy cannot be destroyed, and energy in all of its forms always retains its equivalent amount of mass throughout any transformation to a different type of energy within a system (or translocation into or out of a system). Certain types of *matter* (a different concept) may be created or destroyed, but in all of these processes, the energy and mass associated with such matter remains unchanged in quantity (although type of energy associated with the matter may change form).

In general relativity, mass (and energy) conservation in expanding volumes of space is a complex concept, subject to different definitions, and neither mass nor energy is as strictly and simply conserved as is the case in special relativity and in Minkowski space.

History

An important idea in ancient Greek philosophy was that "Nothing comes from nothing", so that what exists now has always existed: no new matter can come into existence where there was none before. An explicit statement of this, along with the further principle that nothing can pass away into nothing, is found in Empedocles (approx. 490–430 BC): “For it is impossible for anything to come to be from what is not, and it cannot be brought about or heard of that what is should be utterly destroyed.”

A further principle of conservation was stated by Epicurus (341–270 BC) who, describing the nature of the Universe, wrote that “the totality of things was always such as it is now, and always will be”.

Jain philosophy, a non-creationist philosophy based on the teachings of Mahavira (6th century BC), states that the universe and its constituents such as matter cannot be destroyed or created. The Jain text Tattvarthasutra (2nd century AD) states that a substance is permanent, but its modes are characterised by creation and destruction. A principle of the conservation of matter was also stated by Nasīr al-Dīn al-Tūsī (1201–1274). He wrote that “A body of matter cannot disappear completely. It only changes its form, condition, composition, color and other properties and turns into a different complex or elementary matter”.

Mass Conservation in Chemistry

The principle of conservation of mass was first outlined by Mikhail Lomonosov (1711–1765) in 1748. He proved it by experiments—though this is sometimes challenged. Antoine Lavoisier (1743–1794) had expressed these ideas in 1774. Others whose ideas pre-dated the work of Lavoisier include Joseph Black (1728–1799), Henry Cavendish (1731–1810), and Jean Rey (1583–1645).

The conservation of mass was obscure for millennia because of the buoyancy effect of the Earth’s atmosphere on the weight of gases. For example, a piece of wood weighs less after burning; this seemed to suggest that some of its mass disappears, or is transformed or lost. This was not disproved until careful experiments were performed in which chemical reactions such as rusting were allowed to take place in sealed glass ampoules; it was found that the chemical reaction did not change the weight of the sealed container and its contents. The vacuum pump also enabled the weighing of gases using scales.

Once understood, the conservation of mass was of great importance in progressing from alchemy to modern chemistry. Once early chemists realized that chemical substances never disappeared but were only transformed into other substances with the same weight, these scientists could for the first time embark on quantitative studies of the transformations of substances. The idea of mass conservation plus a surmise that certain “elemental substances” also could not be transformed into others by chemical reactions, in turn led to an understanding of chemical elements, as well as the idea that all chemical processes and transformations (such as burning and metabolic reactions) are reactions between invariant amounts or weights of these chemical elements.

Following the pioneering work of Lavoisier the prolonged and exhaustive experiments of Jean Stas supported the strict accuracy of this law in chemical reactions, even though they were carried out with other intentions. His research indicated that in certain reactions the loss or gain could not have been more than from 2 to 4 parts in 100,000. The difference in the accuracy aimed at and attained by Lavoisier on the one hand, and by Morley and Stas on the other, is enormous.

Generalization

In special relativity, the conservation of mass does not apply if the system is open and energy escapes. However, it does continue to apply to totally closed (isolated) systems. If energy cannot escape a system, its mass cannot decrease. In relativity theory, so long as any type of energy is retained within a system, this energy exhibits mass.

Also, mass must be differentiated from matter, since matter may *not* be perfectly conserved in isolated systems, even though mass is always conserved in such systems. However, matter is so nearly conserved in chemistry that violations of matter conservation were not measured until the nuclear age, and the assumption of matter conservation remains an important practical concept in most systems in chemistry and other studies that do not involve the high energies typical of radioactivity and nuclear reactions.

The Mass Associated with Chemical Amounts of Energy is too Small to Measure

The change in mass of certain kinds of open systems where atoms or massive particles are not allowed to escape, but other types of energy (such as light or heat) are allowed to enter or escape, went unnoticed during the 19th century, because the change in mass associated with addition or loss of small quantities of thermal or radiant energy in chemical reactions is very small. (In theory, mass would not change at all for experiments conducted in isolated systems where heat and work were not allowed in or out.)

The theoretical association of all energy with mass was made by Albert Einstein in 1905. However Max Planck pointed out that the change in mass of systems as a result of extraction or addition of chemical energy, as predicted by Einstein's theory, is so small that it could not be measured with available instruments, for example as a test of Einstein's theory. Einstein speculated that the energies associated with newly discovered radioactivity were significant enough, compared with the mass of systems producing them, to enable their mass-change to be measured, once the energy of the reaction had been removed from the system. This later indeed proved to be possible, although it was eventually to be the first artificial nuclear transmutation reaction in 1932, demonstrated by Cockcroft and Walton, that proved the first successful test of Einstein's theory regarding mass-loss with energy-loss.

Mass Conservation Remains Correct if Energy is not Lost

The conservation of relativistic mass implies the viewpoint of a single observer (or the view from a single inertial frame) since changing inertial frames may result in a change of the total energy (relativistic energy) for systems, and this quantity determines the relativistic mass.

The principle that the mass of a system of particles must be equal to the sum of their rest masses, even though true in classical physics, may be false in special relativity. The reason that rest masses cannot be simply added is that this does not take into account other forms of energy, such as kinetic and potential energy, and massless particles such as photons, all of which may (or may not) affect the total mass of systems.

For moving massive particles in a system, examining the rest masses of the various particles also amounts to introducing many different inertial observation frames (which is prohibited if total system energy and momentum are to be conserved), and also when in the rest frame of one particle, this procedure ignores the momenta of other particles, which affect the system mass if the other particles are in motion in this frame.

For the special type of mass called invariant mass, changing the inertial frame of observation for a whole closed system has no effect on the measure of invariant mass of the system, which remains both conserved and invariant (unchanging), even for different observers who view the entire system. Invariant mass is a system combination of energy and momentum, which is invariant for any observer, because in any inertial frame, the energies and momenta of the various particles always add to the same quantity (the momentum may be negative, so the addition amounts to a subtraction). The invariant mass is the relativistic mass of the system when viewed in the center of momentum frame. It is the minimum mass which a system may exhibit, as viewed from all possible inertial frames.

The conservation of both relativistic and invariant mass applies even to systems of particles created by pair production, where energy for new particles may come from kinetic energy of other particles, or from one or more photons as part of a system that includes other particles besides a photon. Again, neither the relativistic nor the invariant mass of totally closed (that is, isolated) systems changes when new particles are created. However, different inertial observers will disagree on the value of this conserved mass, if it is the relativistic mass (i.e., relativistic mass is conserved by not invariant). However, all observers agree on the value of the conserved mass if the mass being measured is the invariant mass (i.e., invariant mass is both conserved and invariant).

The mass-energy equivalence formula gives a different prediction in non-isolated systems, since if energy is allowed to escape a system, both relativistic mass and invariant mass will escape also. In this case, the mass-energy equivalence formula predicts that the *change* in mass of a system is associated with the *change* in its energy due to energy being added or subtracted: $\Delta m = \Delta E / c^2$. This form involving changes was the form in which this famous equation was originally presented by Einstein. In this sense, mass changes in any system are explained simply if the mass of the energy added or removed from the system, are taken into account.

The formula implies that bound systems have an invariant mass (rest mass for the system) less than the sum of their parts, if the binding energy has been allowed to escape the system after the system has been bound. This may happen by converting system potential energy into some other kind of active energy, such as kinetic energy or photons, which easily escape a bound system. The difference in system masses, called a mass defect, is a measure of the binding energy in bound systems – in other words, the energy needed to break the system apart. The greater the mass defect, the larger the binding energy. The binding energy (which itself has mass) must be released (as light or heat) when the parts combine to form the bound system, and this is the reason the mass of the bound system decreases when the energy leaves the system. The total invariant mass is actually conserved, when the mass of the binding energy that has escaped, is taken into account.

Exceptions or Caveats to Mass/Matter Conservation

Matter is not Perfectly Conserved

The principle of *matter* conservation may be considered as an approximate physical law that is true only in the classical sense, without consideration of special relativity and quantum mechanics. It is approximately true except in certain high energy applications.

A particular difficulty with the idea of conservation of "matter" is that "matter" is not a well-defined word scientifically, and when particles that are considered to be "matter" (such as electrons and positrons) are annihilated to make photons (which are often *not* considered matter) then conservation of matter does not take place over time, even within isolated systems. However, matter is conserved to such an extent that matter conservation may be safely assumed in chemical reactions and all situations in which radioactivity and nuclear reactions are not involved.

Even when matter is not conserved, the collection of mass and energy within the system are conserved.

Open Systems and Thermodynamically Closed Systems

Mass is also not generally conserved in *open* systems. Such is the case when various forms of energy are allowed into, or out of, the system. However, again unless radioactivity or nuclear reactions are involved, the amount of energy escaping such systems as heat, work, or electromagnetic radiation is usually too small to be measured as a decrease in system mass.

The law of mass conservation for isolated systems (totally closed to all mass and energy), as viewed over time from any single inertial frame, continues to be true in modern physics. The reason for this is that relativistic equations show that even "massless" particles such as photons still add mass and energy to isolated systems, allowing mass (though not matter) to be strictly conserved in all processes where energy does not escape the system. In relativity, different observers may disagree as to the particular *value* of the conserved mass of a given system, but each observer will agree that this value does not change over time as long as the system is isolated (totally closed to everything).

General Relativity

In general relativity, the total invariant mass of photons in an expanding volume of space will decrease, due to the red shift of such an expansion. The conservation of both mass and energy therefore depends on various corrections made to energy in the theory, due to the changing gravitational potential energy of such systems.

Thermochemistry

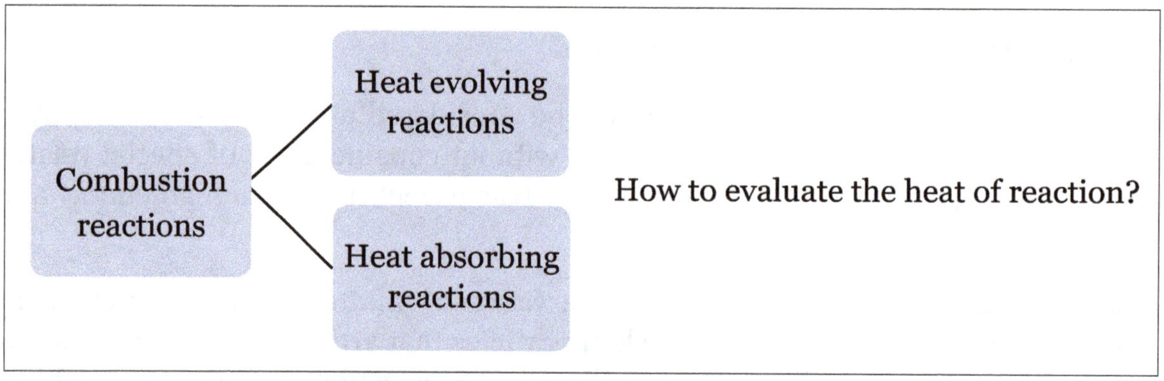

Consider the Burner as Shown Below

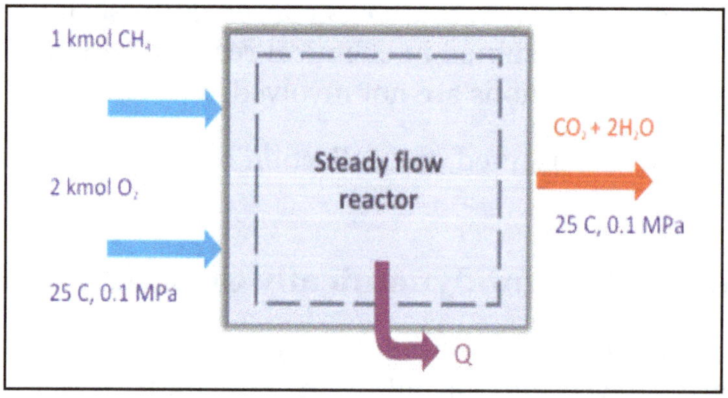

Assumption: (i) Negligible change in K.E. & P.E., (ii) No shaft work,

$$Q = dH = H_p - H_R = \sum_{iP} n_{iP} h_{iP} - \sum_{iR} n_{iR} h_{iR} = \Delta H^{\circ}_{R,298}$$

Where,

H_R – Total enthalpy of reactants; HP – Total enthalpy of products.

n_{iR} – No of moles of ith reactant; niP – No of moles of ith product.

h_{iR} – Enthalpy of formation per unit mole of ith reactant.

h_{iP} – Enthalpy of formation per unit mole of ith product.

$\Delta H^{\circ}_{R,298}$ – Standard heat of reaction.

H_P – Total enthalpy of products.

n_{iP} – No of moles of ith product.

Heat of Reaction Depends on Temperature

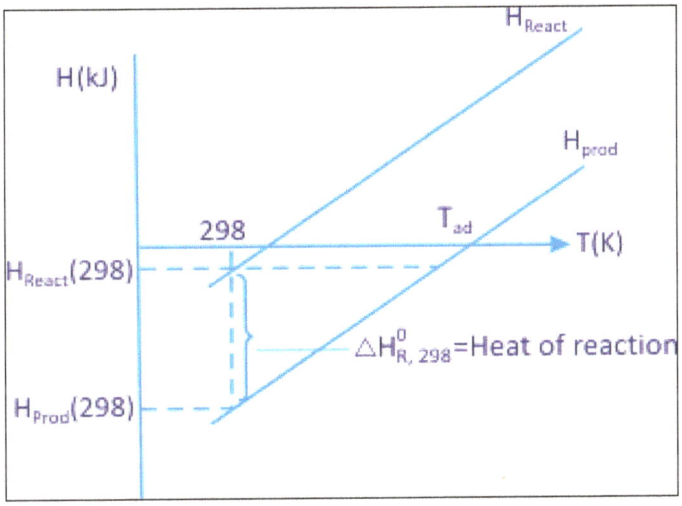

References

- Chris Vuille; Serway, Raymond A.; Faughn, Jerry S. (2009). College physics. Belmont, CA: Brooks/Cole, Cengage Learning. p. 355. ISBN 0-495-38693-6
- What's in a Name? Amount of Substance, Chemical Amount, and Stoichiometric Amount Carmen J. Giunta Journal of Chemical Education 2016 93 (4), 583-586 doi:10.1021/acs.jchemed.5b00690
- "Energy Rules! Energy Conversion and the Laws of Thermodynamics - More About the First and Second Laws". Uwsp.edu. Archived from the original on 5 June 2010. Retrieved 12 September 2010
- Robert D. Whitaker, "An Historical Note on the Conservation of Mass", Journal of Chemical Education, 52, 10, 658-659, Oct 75
- Mahavira is dated 598 BC - 526 BC. Dundas, Paul; John Hinnels ed. (2002). The Jains. London: Routledge. ISBN 0-415-26606-8
- Ben-Naim, A. (2008). A Farewell to Entropy: Statistical Thermodynamics Based on Information, World Scientific, New Jersey, ISBN 978-981-270-706-2

- Smith, J.M.; Van Ness, H.C.; Abbott, M.M. (2005). Introduction to Chemical Engineering Thermodynamics. McGraw Hill. ISBN 0-07-310445-0. OCLC 56491111
- "Conservation of Mass in Chemical Changes"Journal - Chemical Society, London, Vol.64, Part 2 Chemical Society (Great Britain)
- Maas, U.; Pope, S.B. (1992). "Simplifying chemical kinetics: intrinsic low-dimensional manifolds in composition space". Combust. Flame. 88: 239–264. doi:10.1016/0010-2180(92)90034-m
- Bradley, D (2009-06-25). "Combustion and the design of future engine fuels". Proceedings of the Institution of Mechanical Engineers, Part C: Journal of Mechanical Engineering Science. 223 (12): 2751–2765. doi:10.1243/09544062jmes1519
- Valorani, M.; Paolucci, S. (2009). "The G-Scheme: a framework for multi-scale adaptive model reduction". J. Comput. Phys. 228: 4665–4701. Bibcode:2009JCoPh.228.4665V. doi:10.1016/j.jcp.2009.03.011
- Chiavazzo, Eliodoro (2012). "Approximation of slow and fast dynamics in multiscale dynamical systems by the linearized Relaxation Redistribution Method". Journal of Computational Physics. 231: 1751–1765
- Chiavazzo, Eliodoro; Karlin, Ilya; Gorban, Alexander (2010). "The role of thermodynamics in model reduction when using invariant grids". Commun. Comput. Phys. 8: 701–734. doi:10.4208/cicp.030709.210110a
- Kooshkbaghi, Mahdi; Frouzakis, E. Christos; Chiavazzo, Eliodoro; Boulouchos, Konstantinos; Karlin, Ilya (2014). "The global relaxation redistribution method for reduction of combustion kinetics". The Journal of Chemical Physics. 141: 044102. Bibcode:2014JChPh.141d4102K. doi:10.1063/1.4890368

5

Significant Aspects of Automotive Engines

Some of the important aspects of automotive engines are vehicles emissions control, automotive electronics, automatic transmission, manufacturing engineering, g-force, fuel efficiency, safety engineering, engine power and revolutions per minute. This chapter has been carefully written to provide an easy understanding of these facets of automotive engines.

Vehicle Emissions Control

Vehicle emissions control is the study of reducing the motor vehicle emissions—emissions produced by motor vehicles, especially internal combustion engines.

Types of Emissions

Emissions of many air pollutants have been shown to have variety of negative effects on public health and the natural environment. Emissions that are principal pollutants of concern include:

- Hydrocarbons (HC) - A class of burned or partially burned fuel, hydrocarbons are toxins. Hydrocarbons are a major contributor to smog, which can be a major problem in urban areas. Prolonged exposure to hydrocarbons contributes to asthma, liver disease, lung disease, and cancer. Regulations governing hydrocarbons vary according to type of engine and jurisdiction; in some cases, "non-methane hydrocarbons" are regulated, while in other cases, "total hydrocarbons" are regulated. Technology for one application (to meet a non-methane hydrocarbon standard) may not be suitable for use in an application that has to meet a total hydrocarbon standard. Methane is not directly toxic, but is more difficult to break down in a catalytic converter, so in effect a "non-methane hydrocarbon" regulation can be considered easier to meet. Since methane is a greenhouse gas, interest is rising in how to eliminate emissions of it. HC emissions can come not only from a vehicle's engine buts also directly from the fuel tank and lines, 24 hours a day, even when the engine is off; the complex system of fuel vent lines and a charcoal canister is meant to collect and contain fuel vapors and route them either back to the fuel tank or, after the engine is started and warmed up, into the air intake to be burned in the engine.

- Carbon monoxide (CO) - A product of incomplete combustion, inhaled carbon monoxide reduces the blood's ability to carry oxygen; overexposure (carbon monoxide poisoning) may be fatal. (Carbon monoxide persistently binds to hemoglobin, the oxygen-carrying chemical in red blood cells, where oxygen (O_2) would temporarily bind; the bonding of CO excludes O_2 and also reduces the ability of the hemoglobin to release already-bound oxygen, on both counts rendering the red blood cells ineffective. Recovery is by the slow release of bound CO and the body's production of new hemoglobin—a healing process—so full recovery from moderate to severe [but nonfatal] CO poisoning takes hours or days. Removing a person from a CO-poisoned atmosphere to fresh air stops the injury but does not yield prompt recovery, unlike the case where a person is removing from an asphyxiating atmosphere [i.e. one deficient in oxygen]. Toxic effects delayed by days are also common.)

- NO_x - Generated when nitrogen in the air reacts with oxygen at the high temperature and pressure inside the engine. NO_x is a precursor to smog and acid rain. NO_x is the sum of NO and NO_2. NO_2 is extremely reactive. NO_x production is increased when an engine runs at its most efficient (i.e. hottest) operating point, so there tends to be a natural tradeoff between efficiency and control of NO_x emissions.
- Particulate matter – Soot or smoke made up of particles in the micrometre size range: Particulate matter causes negative health effects, including but not limited to respiratory disease and cancer. Very fine particulate matter has been linked to cardiovascular disease.
- Sulfur oxide (SO_x) - A general term for oxides of sulfur, which are emitted from motor vehicles burning fuel containing sulfur. Reducing the level of fuel sulfur reduces the level of Sulfur oxide emitted from the tailpipe.
- Volatile organic compounds (VOCs) - Organic compounds which typically have a boiling point less than or equal to 250 °C; for example chlorofluorocarbons (CFCs) and formaldehyde. Volatile organic compounds are a subsection of Hydrocarbons that are mentioned separately because of their dangers to public health.

History

Throughout the 1950s and 1960s, various federal, state and local governments in the United States conducted studies into the numerous sources of air pollution. These studies ultimately attributed a significant portion of air pollution to the automobile, and concluded air pollution is not bounded by local political boundaries. At that time, such minimal emission control regulations as existed in the U.S. were promulgated at the municipal or, occasionally, the state level. The ineffective local regulations were gradually supplanted by more comprehensive state and federal regulations. By 1967 the State of California created the California Air Resources Board, and in 1970, the federal United States Environmental Protection Agency (EPA) was established. Both agencies, as well as other state agencies, now create and enforce emission regulations for automobiles in the United States. Similar agencies and regulations were contemporaneously developed and implemented in Canada, Western Europe, Australia, and Japan.

The first effort at controlling pollution from automobiles was the PCV (positive crankcase ventilation) system. This draws crankcase fumes heavy in unburned hydrocarbons — a precursor to photochemical smog — into the engine's intake tract so they are burned rather than released unburned from the crankcase into the atmosphere. Positive crankcase ventilation was first installed on a widespread basis by law on all new 1961-model cars first sold in California. The following year, New York required it. By 1964, most new cars sold in the U.S. were so equipped, and PCV quickly became standard equipment on all vehicles worldwide.

The first legislated exhaust (tailpipe) emission standards were promulgated by the State of

California for 1966 model year for cars sold in that state, followed by the United States as a whole in model year 1968. Also in 1966, the first emission test cycle was enacted in the State of California measuring tailpipe emissions in PPM (parts per million). The standards were progressively tightened year by year, as mandated by the EPA.

By the 1974 model year, the emission standards had tightened such that the de-tuning techniques used to meet them were seriously reducing engine efficiency and thus increasing fuel usage. The new emission standards for 1975 model year, as well as the increase in fuel usage, forced the invention of the catalytic converter for after-treatment of the exhaust gas. This was not possible with existing leaded gasoline, because the lead residue contaminated the platinum catalyst. In 1972, General Motors proposed to the American Petroleum Institute the elimination of leaded fuels for 1975 and later model year cars. The production and distribution of unleaded fuel was a major challenge, but it was completed successfully in time for the 1975 model year cars. All modern cars are now equipped with catalytic converters and leaded fuel is nearly impossible to buy in most First World countries.

Regulatory Agencies

The agencies charged with regulating exhaust emissions vary from jurisdiction to jurisdiction, even in the same country. For example, in the United States, overall responsibility belongs to the EPA, but due to special requirements of the State of California, emissions in California are regulated by the Air Resources Board. In Texas, the Texas Railroad Commission is responsible for regulating emissions from LPG-fueled rich burn engines (but not gasoline-fueled rich burn engines).

North America

- California Air Resources Board - California, United States (most sources).
- Environment Canada - Canada (most sources).
- Environmental Protection Agency - United States (most sources).
- Texas Railroad Commission - Texas, United States (LPG-fueled engines only).
- Transport Canada - Canada (trains and ships).

Europe

The European Union has control over regulation of emissions in EU member states; however, many member states have their own government bodies to enforce and implement these regulations in their respective countries. In short, the EU forms the policy (by setting limits such as the European emission standard) and the member states decide how to best implement it in their own country.

United Kingdom

In the United Kingdom, matters concerning environmental policy are what is known as "devolved powers" which means, each of the constituent countries deals with it separately through their own government bodies set up to deal with environmental issues in their respective country:

- Environment Agency - England and Wales.
- Scottish Environment Protection Agency (SEPA) - Scotland.
- Department of the Environment - Northern Ireland.

However, many UK-wide policies are handled by the Department of the Environment Food and Rural Affairs (DEFRA) and they are still subject to EU regulations.

Emissions Control

Engine efficiency has been steadily improved with improved engine design, more precise ignition timing and electronic ignition, more precise fuel metering, and computerized engine management.

Advances in engine and vehicle technology continually reduce the toxicity of exhaust leaving the engine, but these alone have generally been proved insufficient to meet emissions goals. Therefore, technologies to detoxify the exhaust are an essential part of emissions control.

Air Injection

One of the first-developed exhaust emission control systems is secondary air injection. Originally, this system was used to inject air into the engine's exhaust ports to provide oxygen so unburned and partially burned hydrocarbons in the exhaust would finish burning. Air injection is now used to support the catalytic converter's oxidation reaction, and to reduce emissions when an engine is started from cold. After a cold start, an engine needs an air-fuel mixture richer than what it needs at operating temperature, and the catalytic converter does not function efficiently until it has reached its own operating temperature. The air injected upstream of the converter supports combustion in the exhaust headpipe, which speeds catalyst warmup and reduces the amount of unburned hydrocarbon emitted from the tailpipe.

Exhaust Gas Recirculation

In the United States and Canada, many engines in 1973 and newer vehicles (1972 and newer in California) have a system that routes a metered amount of exhaust into the intake tract under particular operating conditions. Exhaust neither burns nor supports combustion, so it dilutes the air/fuel charge to reduce peak combustion chamber temperatures. This, in turn, reduces the formation of NO_x.

Catalytic Converter

The catalytic converter is a device placed in the exhaust pipe, which converts hydrocarbons, carbon monoxide, and NO_x into less harmful gases by using a combination of platinum, palladium and rhodium as catalysts.

There are two types of catalytic converter, a two-way and a three-way converter. Two-way converters were common until the 1980s, when three-way converters replaced them on most automobile engines.

Evaporative Emissions Control

EVAP may also refer to Evaporation.

Evaporative emissions are the result of gasoline vapors escaping from the vehicle's fuel system. Since 1971, all U.S. vehicles have had fully sealed fuel systems that do not vent directly to the atmosphere; mandates for systems of this type appeared contemporaneously in other jurisdictions. In a typical system, vapors from the fuel tank and carburetor bowl vent (on carbureted vehicles) are ducted to canisters containing activated carbon. The vapors are adsorbed within the canister, and during certain engine operational modes fresh air is drawn through the canister, pulling the vapor into the engine, where it burns.

Remote Sensing Emission Testing - Field Studies

Some US states are also using a technology developed by Dr. Donald H. Stedman of the University of Denver, which uses infra-red and ultraviolet light to detect emissions while vehicles pass by on public roads, thus eliminating the need for owners to go to a test center. Stedman's invisible light flash detection of exhaust gases is commonly used in metropolitan areas, is offered by the US-Swedish company OPUS Inspection and becoming more broadly known in Europe.

Use of Emission Test Data

Emission test results from individual vehicles are in many cases compiled to evaluate the emissions performance of various classes of vehicles, the efficacy of the testing program and of various other emission-related regulations (such as changes to fuel formulations) and to model the effects of auto emissions on public health and the environment. For example, the Environmental Working Group used California ASM emissions data to create an "Auto Asthma Index" that rates vehicle models according to emissions of hydrocarbons and nitrogen oxides, chemical precursors to photochemical smog.

Automotive Electronics

Automotive electronics are any electrically-generated systems used in road vehicles, such as: carputers, telematics, in-car entertainment systems, etc.

Automotive electronics originated from the need to control engines. The first electronic pieces were used to control engine functions and were referred to as engine control units (ECU). As electronic controls began to be used for more automotive applications, the acronym ECU took on the more general meaning of "electronic control unit", and then specific ECU's were developed. Now, ECU's are modular. Two types include engine control modules (ECM) or transmission control modules (TCM).

A modern car may have up to 100 ECU's and a commercial vehicle up to 40.

Automotive electronics or automotive embedded systems are distributed systems, and according to different domains in the automotive field, they can be classified into:

- Engine electronics

- Transmission electronics
- Chassis electronics
- Active safety
- Driver assistance
- Passenger comfort
- Entertainment systems

Engine Electronics

One of the most demanding electronic parts of an automobile is the engine control unit. Engine controls demand one of the highest real time deadlines, as the engine itself is a very fast and complex part of the automobile. Of all the electronics in any car the computing power of the engine control unit is the highest, typically a 32-bit processor.

It controls such things as:

In a diesel engine:

- Fuel injection rate
- Emission control, NOx control
- Regeneration of oxidation catalytic converter
- Turbocharger control
- Cooling system control
- Throttle control

In a gasoline engine:

- Lambda control
- OBD (On-Board Diagnostics)
- Cooling system control
- Ignition system control
- Lubrication system control (only a few have electronic control)
- Fuel injection rate control
- Throttle control

Many more engine parameters are actively monitored and controlled in real-time. There are about 20 to 50 that measure pressure, temperature, flow, engine speed, oxygen level and NOx level plus other parameters at different points within the engine. All these sensor signals are sent to the

ECU, which has the logic circuits to do the actual controlling. The ECU output is connected to different actuators for the throttle valve, EGR valve, rack (in VGTs), fuel injector (using a pulse-width modulated signal), dosing injector and more. There are about 20 to 30 actuators in all.

Transmission Electronics

These control the transmission system, mainly the shifting of the gears for better shift comfort and to lower torque interrupt while shifting. Automatic transmissions use controls for their operation, and also many semi-automatic transmissions having a fully automatic clutch or a semi-auto clutch (declutching only). The engine control unit and the transmission control exchange messages, sensor signals and control signals for their operation.

Chassis Electronics

The chassis system has lot of sub-systems which monitor various parameters and are actively controlled:

- ABS - Anti-lock Braking System
- TCS – Traction Control System
- EBD – Electronic Brake Distribution
- ESP – Electronic Stability Program

Passive Safety

These systems are always ready to act when there is a collision in progress or to prevent it when it senses a dangerous situation:

- Air bags
- Hill descent control
- Emergency brake assist system

Driver Assistance

- Lane assist system
- Speed assist system
- Blind spot detection
- Park assist system
- Adaptive cruise control system

Passenger Comfort

- Automatic climate control
- Electronic seat adjustment with memory

- Automatic wipers
- Automatic headlamps - adjusts beam automatically
- Automatic cooling - temperature adjustment

Infotainment Systems

- Navigation system
- Vehicle audio
- Information access

All of the above systems forms an infotainment system. Developmental methods for these systems vary according to each manufacturer. Different tools are used for both hardware and software development.

Functional Safety Requirements

In order to minimize the risk of dangerous failures, safety related electronic systems have to be developed following the applicable product liability requirements. Disregard for, or inadequate application of these standards can lead to not only personal injuries, but also severe legal and economic consequences such as product cancellations or recalls.

The IEC 61508 standard, generally applicable to electrical/electronic/programmable safety-related products, is only partially adequate for automotive-development requirements. Consequently, for the automotive industry, this standard is replaced by the existing ISO 26262, currently released as a Final Draft International Standard (FDIS). ISO/DIS 26262 describes the entire product life-cycle of safety related electrical/electronic systems for road vehicles. It has been published as an international standard in its final version in November 2011. The implementation of this new standard will result in modifications and various innovations in the automobile electronics development process, as it covers the complete product life-cycle from the concept phase until its decommissioning.

Automatic Transmission

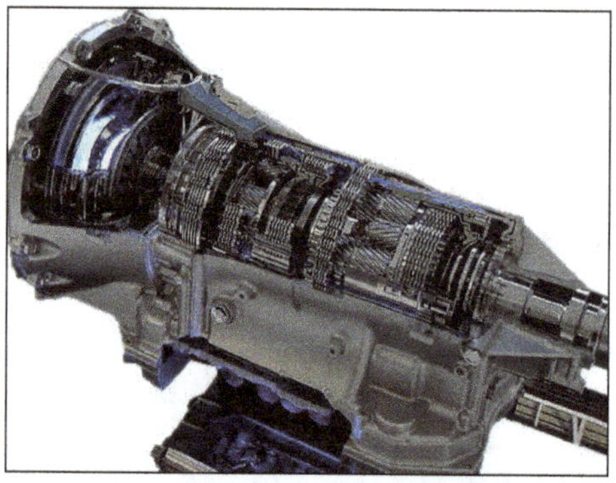

An 8-gear automatic transmission.

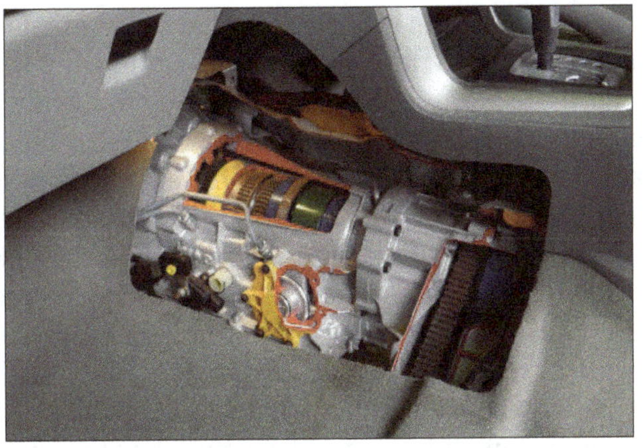

Cutaway showing the typical positioning of an automatic transmission from the interior of an automobile.

An automatic transmission, also called auto, self-shifting transmission, n-speed automatic (where n is its number of forward gear ratios), or AT, is a type of motor vehicle transmission that can automatically change gear ratios as the vehicle moves, freeing the driver from having to shift gears manually. Like other transmission systems on vehicles, it allows an internal combustion engine, best suited to run at a relatively high rotational speed, to provide a range of speed and torque outputs necessary for vehicular travel. The number of forward gear ratios is often expressed for manual transmissions as well (e.g., 6-speed manual).

The most popular form found in automobiles is the hydraulic automatic transmission. Similar but larger devices are also used for heavy-duty commercial and industrial vehicles and equipment. This system uses a fluid coupling in place of a friction clutch, and accomplishes gear changes by hydraulically locking and unlocking a system of planetary gears. These systems have a defined set of gear ranges, often with a parking pawl that locks the output shaft of the transmission to keep the vehicle from rolling either forward or backward. Some machines with limited speed ranges or fixed engine speeds, such as some forklifts and lawn mowers, only use a torque converter to provide a variable gearing of the engine to the wheels.

Besides the traditional hydraulic automatic transmissions, there are also other types of automated transmissions, such as a continuously variable transmission (CVT) and semi-automatic transmissions, that free the driver from having to shift gears manually, by using the transmission's computer to change gear, if for example the driver were redlining the engine. Despite superficial similarity to other transmissions, traditional automatic transmissions differ significantly in internal operation and driver's feel from semi-automatics and CVTs. In contrast to conventional automatic transmissions, a CVT uses a belt or other torque transmission scheme to allow an "infinite" number of gear ratios instead of a fixed number of gear ratios. A semi-automatic retains a clutch like a manual transmission, but controls the clutch through electrohydraulic means. The ability to shift gears manually, often via paddle shifters, can also be found on certain automated transmissions (manumatics such as Tiptronic), semi-automatics (BMW SMG, VW Group DSG), and CVTs (such as Lineartronic).

The obvious advantage of an automatic transmission to the driver is the lack of a clutch pedal and manual shift pattern in normal driving. This allows the driver to operate the car with as few as two limbs (possibly using assist devices to position controls within reach of usable limbs), allowing amputees and other disabled individuals to drive. The lack of manual shifting also reduces the attention and workload required inside the cabin, such as monitoring the tachometer and taking a hand off the wheel to move the shifter, allowing the driver to ideally keep both hands on the wheel at all times and to focus more on the road. Control of the car at low speeds is often easier with an automatic than a manual, due to a side effect of the clutchless fluid-coupling design called "creep" that causes the car to want to move while in a driving gear, even at idle. The primary disadvantage of the most popular hydraulic designs is reduced mechanical efficiency of the power transfer between engine and drivetrain, due to the fluid coupling connecting the engine to the gearbox. This can result in lower power/torque ratings for automatics compared to manuals with the same engine specs, as well as reduced fuel efficiency in city driving as the engine must maintain idle against the resistance of the fluid coupling. Advances in transmission and coupler design have narrowed this gap considerably, but clutch-based transmissions (manual or semi-automatic) are still preferred in sport-tuned trim levels of various production cars, as well as in many auto racing leagues.

The automatic transmission was invented in 1921 by Alfred Horner Munro of Regina, Saskatchewan, Canada, and patented under Canadian patent CA 235757 in 1923. (Munro obtained UK patent GB215669 215,669 for his invention in 1924 and US patent 1,613,525 on 4 January 1927). Being a steam engineer, Munro designed his device to use compressed air rather than hydraulic fluid, and so it lacked power and never found commercial application. The first automatic transmission using hydraulic fluid may have been developed in 1932 by two Brazilian engineers, José Braz Araripe and Fernando Lehly Lemos; subsequently the prototype and plans were sold to General Motors who introduced it in the 1940 Oldsmobile as the "Hydra-Matic" transmission. They were incorporated into GM-built tanks during World War II and, after the war, GM marketed them as being "battle-tested." However, a Wall Street Journal article credits ZF Friedrichshafen with the invention, occurring shortly after World War I. ZF's origins were in manufacturing gears for airship engines beginning in 1915; the company was founded by Ferdinand von Zeppelin.

History

Modern automatic transmissions can trace their origins to an early "horseless carriage" gearbox that was developed in 1904 by the Sturtevant brothers of Boston, Massachusetts. This unit had two forward speeds, the ratio change being brought about by flyweights that were driven by the engine. At higher engine speeds, high gear was engaged. As the vehicle slowed down and engine RPM decreased, the gearbox would shift back to low. Unfortunately, the metallurgy of the time wasn't up to the task, and owing to the abruptness of the gear change, the transmission would often fail without warning.

The next significant phase in the automatic transmission's development occurred in 1908 with the introduction of Henry Ford's remarkable Model T. The Model T, in addition to being cheap and reliable by the standards of the day, featured a simple, two speed plus reverse planetary transmission whose operation was manually controlled by the driver using pedals. The pedals actuated the transmission's friction elements (bands and clutches) to select the desired gear. In some respects, this type of transmission was less demanding of the driver's skills than the contemporary, unsynchronized manual transmission, but still required that the driver know when to make a shift, as well as how to get the car off to a smooth start.

In 1934, both REO and General Motors developed semi-automatic transmissions that were less difficult to operate than a fully manual unit. These designs, however, continued to use a clutch to engage the engine with the transmission. The General Motors unit, dubbed the "Automatic Safety Transmission," was notable in that it employed a power-shifting planetary gearbox that was hydraulically controlled and was sensitive to road speed, anticipating future development.

Parallel to the development in the 1930s of an automatically shifting gearbox was Chrysler's work on adapting the fluid coupling to automotive use. Invented early in the 20th century, the fluid coupling was the answer to the question of how to avoid stalling the engine when the vehicle was stopped with the transmission in gear. Chrysler itself never used the fluid coupling with any of its automatic transmissions, but did use it in conjunction with a hybrid manual transmission called "Fluid Drive" (the similar Hy-Drive used a torque converter). These developments in automatic gearbox and fluid coupling technology eventually culminated in the introduction in 1939 of the General Motors Hydra-Matic, the world's first mass-produced automatic transmission.

Available as an option on 1940 Oldsmobiles and later Cadillacs, the Hydra-Matic combined a fluid coupling with three hydraulically controlled planetary gearsets to produce four forward speeds plus reverse. The transmission was sensitive to engine throttle position and road speed, producing fully automatic up- and down-shifting that varied according to operating conditions.

The Hydra-Matic was subsequently adopted by Cadillac and Pontiac, and was sold to various other automakers, including Bentley, Hudson, Kaiser, Nash, and Rolls-Royce. It also found use during World War II in some military vehicles. From 1950 to 1954, Lincoln cars were also available with the Hydra-Matic. Mercedes-Benz subsequently devised a four-speed fluid coupling transmission that was similar in principle to the Hydra-Matic, but of a different design.

Interestingly, the original Hydra-Matic incorporated two features which are widely emulated in today's transmissions. The Hydra-Matic's ratio spread through the four gears produced excellent "step-off" and acceleration in first, good spacing of intermediate gears, and the effect of an overdrive in fourth, by virtue of the low numerical rear axle ratio used in the vehicles of the time. In addition, in third and fourth gear, the fluid coupling only handled a portion of the engine's torque, resulting in a high degree of efficiency. In this respect, the transmission's behavior was similar to modern units incorporating a lock-up torque converter.

In 1956, GM introduced the "Jetaway" Hydra-Matic, which was different in design than the older model. Addressing the issue of shift quality, which was an ongoing problem with the original Hydra-Matic, the new transmission utilized two fluid couplings, the primary one that linked the transmission to the engine, and a secondary one that replaced the clutch assembly that controlled the forward gearset in the original. The result was much smoother shifting, especially from first to second gear, but with a loss in efficiency and an increase in complexity. Another innovation for this new style Hydra-Matic was the appearance of a Park position on the selector. The original Hydra-Matic, which continued in production until the mid-1960s, still used the reverse position for parking pawl engagement.

The first torque converter automatic, Buick's Dynaflow, was introduced for the 1948 model year.

It was followed by Packard's Ultramatic in mid-1949 and Chevrolet's Powerglide for the 1950 model year. Each of these transmissions had only two forward speeds, relying on the converter for additional torque multiplication. In the early 1950s, BorgWarner developed a series of three-speed torque converter automatics for American Motors, Ford Motor Company, Studebaker, and several other manufacturers in the US and other countries. Chrysler was late in developing its own true automatic, introducing the two-speed torque converter PowerFlite in 1953, and the three-speed TorqueFlite in 1956. The latter was the first to utilize the Simpson compound planetary gearset.

General Motors produced multiple-turbine torque converters from 1954 to 1961. These included the Twin-Turbine Dynaflow and the triple-turbine Turboglide transmissions. The shifting took place in the torque converter, rather than through pressure valves and changes in planetary gear connections. Each turbine was connected to the drive shaft through a different gear train. These phased from one ratio to another according to demand, rather than shifting. The Turboglide actually had two speed ratios in reverse, with one of the turbines rotating backwards.

By the late 1960s, most of the fluid-coupling four-speed and two-speed transmissions had disappeared in favor of three-speed units with torque converters. Also around this time, whale oil was removed from automatic transmission fluid. By the early 1980s, these were being supplemented and eventually replaced by overdrive-equipped transmissions providing four or more forward speeds. Many transmissions also adopted the lock-up torque converter (a mechanical clutch locking the torque converter pump and turbine together to eliminate slip at cruising speed) to improve fuel economy.

As computerized engine control units (ECUs) became more capable, much of the logic built into the transmission's valve body was offloaded to the ECU. Some manufacturers use a separate computer dedicated to the transmission called a transmission control unit (TCU), also known as the transmission control module (TCM), which share information with the engine management computer. In this case, solenoids turned on and off by the computer control shift patterns and gear ratios, rather than the spring-loaded valves in the valve body. This allows for more precise control of shift points, shift quality, lower shift times, and (on some newer cars) semi-automatic control, where the driver tells the computer when to shift. The result is an impressive combination of efficiency and smoothness. Some computers even identify the driver's style and adapt to best suit it.

ZF Friedrichshafen and BMW were responsible for introducing the first six-speed (the ZF 6HP26 in the 2002 BMW E65 7-Series). Mercedes-Benz's 7G-Tronic was the first seven-speed in 2003, with Toyota introducing an eight-speed in 2007 on the Lexus LS 460. Derived from the 7G-Tronic, Mercedes-Benz unveiled a semi-automatic transmission with the torque converter replaced with a wet multi clutch called the AMG SPEEDSHIFT MCT. The 2014 Jeep Cherokee has the world's first nine-speed automatic transmission for a passenger vehicle to market.

Parts and Operation

Hydraulic Automatic Transmissions

The predominant form of automatic transmission is hydraulically operated; using a fluid coupling or torque converter, and a set of planetary gearsets to provide a range of gear ratios.

A cutaway of an 8-speed ZF 8HP showing the major stages of a hydraulic automatic transmission: the torque converter (left), the planetary gearsets and clutch plates (center), as well as hydraulic and electronic controls (bottom).

Hydraulic automatic transmissions consist of three major components:

Torque Converter

A type of fluid coupling, hydraulically connecting the engine to the transmission. This takes the place of a friction clutch in a manual transmission. It transmits and decouples the engine power to the planetary gears, allowing the vehicle to come to stop with the engine still running without stalling.

A torque converter differs from a fluid coupling, in that it provides a variable amount of torque multiplication at low engine speeds, increasing breakaway acceleration. A fluid coupling works well when both the impeller and turbine are rotating at similar speeds, but it is very inefficient at initial acceleration, where rotational speeds are very different. This torque multiplication is accomplished with a third member in the coupling assembly known as the stator, which acts to modify the fluid flow depending on the relative rotational speeds of the impeller and turbine. The stator itself does not rotate, but its vanes are so shaped that when the impeller (which is driven by the engine) is rotating at a high speed and the turbine (which receives the transmitted power) is spinning at a low speed, the fluid flow hits the vanes of the turbine in a way that multiplies the torque being applied. This causes the turbine to begin spinning faster as the vehicle accelerates (ideally), and as the relative rotational speeds equalize, the torque multiplication diminishes. Once the impeller and turbine are rotating within 10% of each other's speed, the stator ceases to function and the torque converter acts as a simple fluid coupling.

Planetary Gears Train

Consisting of planetary gear sets as well as clutches and bands. These are the mechanical systems that provide the various gear ratios, altering the speed of rotation of the output shaft depending on which planetary gears are locked.

To effect gear changes, one of two types of clutches or bands are used to hold a particular member of the planetary gearset motionless, while allowing another member to rotate, thereby transmitting torque and producing gear reductions or overdrive ratios. These clutches are actuated by the valve body, their sequence controlled by the transmission's internal programming. Principally, a type of device known as a sprag or roller clutch is used for routine upshifts/downshifts. Operating much as a ratchet, it transmits torque only in one direction, free-wheeling or "overrunning" in the other. The advantage of this type of clutch is that it eliminates the sensitivity of timing a simultaneous clutch release/apply on two planetaries, simply "taking up" the drivetrain load when actuated, and releasing automatically when the next gear's sprag clutch assumes the torque transfer. The bands come into play for manually selected gears, such as low range or reverse, and operate on the planetary drum's circumference. Bands are not applied when drive/overdrive range is selected, the torque being transmitted by the sprag clutches instead. Bands are used for braking; the GM Turbo-Hydramatics incorporated this.

Hydraulic Controls

Uses special transmission fluid sent under pressure by an oil pump to control various clutches and bands modifying the speed of the output depending on the vehicle's running condition.

The pump is typically a gear pump mounted between the torque converter and the planetary gearset. It draws transmission fluid from a sump and pressurizes it, which is needed for transmission components to operate. The input for the pump is connected to the torque converter housing, which in turn is bolted to the engine's flexplate, so the pump provides pressure whenever the engine is running and there is enough transmission fluid, but the disadvantage is that when the engine is not running, no oil pressure is available to operate the main components of the transmission, and is thus impossible to push-start a vehicle equipped with an automatic transmission. Early automatic transmissions also had a rear pump for towing purposes, ensuring the lubrication of the rear-end components.

The governor is connected to the output shaft and regulates the hydraulic pressure depending on the vehicle speed. The engine load is monitored either by a throttle cable or a vacuum modulator. The valve body is the hydraulic control center that receives pressurized fluid from the main pump operated by the fluid coupling/torque converter. The pressure coming from this pump is regulated and used to run a network of spring-loaded valves, check balls and servo pistons. The valves use the pump pressure and the pressure from a centrifugal governor on the output side (as well as hydraulic signals from the range selector valves and the throttle valve or modulator) to control which ratio is selected on the gearset; as the vehicle and engine change speed, the difference between the pressures changes, causing different sets of valves to open and close. The hydraulic pressure controlled by these valves drives the various clutch and brake band actuators, thereby controlling the operation of the planetary gearset to select the optimum gear ratio for the current operating conditions. However, in many modern automatic transmissions, the valves are controlled by electro-mechanical servos which are controlled by the electronic engine control unit (ECU) or a separate transmission control unit (TCU, also known as transmission control module (TCM).

The hydraulic & lubricating oil, called automatic transmission fluid (ATF), provides lubrication, corrosion prevention, and a hydraulic medium to convey mechanical power (for the operation of the transmission). Primarily made from refined petroleum, and processed to provide properties that promote smooth power transmission and increase service life, the ATF is one of the few parts of the automatic transmission that needs routine service as the vehicle ages.

The multitude of parts, along with the complex design of the valve body, originally made hydraulic automatic transmissions much more complicated (and expensive) to build and repair than manual transmissions. In most cars (except US family, luxury, sport-utility vehicle, and minivan models) they have usually been extra-cost options for this reason. Mass manufacturing and decades of improvement have reduced this cost gap.

In some modern cars, computers use sensors on the engine to detect throttle position, vehicle speed, engine speed, engine load, etc. to control the exact shift point. The computer transmits the information via solenoids that redirect the fluid the appropriate clutch or servo to control shifting.

Continuously Variable Transmissions

A fundamentally different type of automatic transmission is the continuously variable transmission or CVT, which can smoothly and steplessly alter its gear ratio by varying the diameter of a pair of belt or chain-linked pulleys, wheels or cones. Some continuously variable transmissions use a hydrostatic drive — consisting of a variable displacement pump and a hydraulic motor — to transmit power without gears. Some early forms, such as the Hall system (which dates back to 1896), used a fixed displacement pump and a variable displacement motor, and were designed to provide robust variable transmission for early commercial heavy motor vehicles. CVT designs are usually as fuel efficient as manual transmissions in city driving, but early designs lose efficiency as engine speed increases.

A slightly different approach to CVT is the concept of toroidal CVT or infinitely variable transmission (IVT). These concepts provide zero and reverse gear ratios.

E-CVT

Some hybrid vehicles, notably those of Toyota, Lexus and Ford Motor Company, have an electronically controlled CVT (E-CVT). In this system, the transmission has fixed gears, but the ratio of wheel-speed to engine-speed can be continuously varied by controlling the speed of the third input to a differential using motor-generators.

Automatic Transmission Modes

Most automatic transmissions offer the driver a certain amount of manual control over the transmission's shifts.

Conventionally, in order to select the transmission operating mode, the driver moves a selection lever located either on the steering column or on the floor (as with a manual on the floor, except that automatic selectors on the floor do not move in the same type of pattern as manual levers do). In order to select modes, or to manually select specific gear ratios, the driver must push a button in (called the shift-lock button) or pull the handle (only on column mounted shifters) out. Some vehicles position selector buttons for each mode on the cockpit instead, freeing up space on the central console.

Vehicles conforming to US Government standards must have the modes ordered P-R-N-D-L (left to right, top to bottom, or clockwise). Previously, quadrant-selected automatic transmissions often used a P-N-D-L-R layout, or similar. Such a pattern led to a number of deaths and injuries owing to driver error causing unintentional gear selection, as well as the danger of having a selector (when worn) jump into reverse from low gear during engine braking maneuvers.

A floor selection lever in a 1992 Ford Escort showing the P-R-N-[D]-D-L modes as well as the shift lock button on the top of the lever.

Depending on the model and make of the transmission, these controls can take several forms. However most include the following:

Park (P):

This selection mechanically locks the output shaft of transmission, restricting the vehicle from moving in any direction. A parking pawl prevents the transmission from rotating, and therefore the vehicle from moving. However, it should be noted that the vehicle's non-driven wheels are still free to rotate, and the driven wheels may still rotate individually (because of the differential). For this reason, it is recommended to use the hand brake (parking brake) because this actually locks (in most cases) the wheels and prevents them from moving. It is typical of front-wheel-drive vehicles for the parking brake to lock the rear (non-driving) wheels, so use of both the parking brake and the transmission park lock provides the greatest security against unintended movement on slopes. This also increases the life of the transmission and the park pin mechanism, because parking on an incline with the transmission in park without the parking brake engaged will cause undue stress on the parking pin, and may even prevent the pin from releasing. A hand brake should also prevent the car from moving if a worn selector accidentally drops into reverse gear while idling.

A car should be allowed to come to a complete stop before setting the transmission into park to prevent damage. Usually, Park (P) is one of only two selections in which the car's engine can be started, the other being Neutral (N). This is typically achieved via a normally open inhibitor switch (sometimes called a "neutral safety switch") wired in series with the starter motor engagement circuit, which is closed when P or N is selected, completing the circuit (when the key is turned to the start position). In many modern cars and trucks, the driver must have the foot brake applied before the transmission can be taken out of park. The Park position is omitted on buses/coaches (and some road tractors) with automatic transmission (on which a parking pawl is not practical), which must instead be placed in neutral with the air-operated parking brakes set.

Reverse (R):

This engages reverse gear within the transmission, permitting the vehicle to be driven backward, and operates a switch to turn on the white backup lights for improved visibility (the switch may also activate a beeper on delivery trucks or other large vehicles to audibly warn other drivers and nearby pedestrians of the driver's reverse movement). To select reverse in most transmissions, the driver must come to a complete stop, depress the shift-lock button (or move the shift lever toward the driver in a column shifter, or move the shifter sideways along a notched channel in a console shifter) and select reverse. The driver should avoid engaging reverse while the vehicle is moving forwards, and likewise avoid engaging any forward gear while travelling backwards. On transmissions with a torque converter, doing so at very low speed (walking pace) is not harmful, but causes unnecessary wear on clutches and bands, and a sudden deceleration that not only is uncomfortable, but also uncontrollable since the brakes and the throttle contribute in the same direction. This sudden acceleration, or jerk, can still be felt when engaging the gear at standstill, but the driver normally suppresses this by holding the brakes. Travelling slowly in the right direction while engaging the gear minimizes the jerk further, which is actually beneficial to the wearing parts of the transmission. Electronically controlled transmissions may behave differently, as engaging a gear at speed is essentially undefined behaviour. Some modern transmissions have a safety mechanism that will resist putting the car in reverse when the vehicle is moving forward; such a mechanism may consist of a solenoid-controlled physical barrier on either side of the reverse position, electronically engaged by a switch on the brake pedal, so that the brake pedal needs to be depressed in order to allow the selection of reverse. Some electronic transmissions prevent or delay engagement of reverse gear altogether while the car is moving.

Some shifters with a shift button allow the driver to freely move the shifter from R to N or D without actually depressing the button. However, the driver cannot shift back to R without depressing the shift button, to prevent accidental shifting which could damage the transmission, especially at high speeds.

Neutral / No gear (N):

This disengages all gear trains within the transmission, effectively disconnecting the transmission from the driven wheels, allowing the vehicle to coast freely under its own weight and gain momentum without the motive force from the engine. Coasting in idle down long grades (where law permits) should be avoided, though, as the transmission's lubrication pump is driven by non-idle engine RPMs. Similarly, emergency towing with an automatic transmission in neutral should be a last resort. Manufacturers understand emergency situations and list limitations of towing a vehicle in neutral (usually not to exceed 55 mph and 50 miles). This is the only other selection in which the vehicle's engine may be started.

Drive (D):

This position allows the transmission to engage the full range of available forward gear ratios, allowing the vehicle to move forward and accelerate through its range of gears. The number of gear ratios within the transmission depends on the model, but they initially ranged from three (predominant before the 1990s), to four and five speeds (losing popularity to six-speed autos). Six-speed automatic transmissions are probably the most common offering in cars and trucks from 2010 in carmakers as Toyota, GM and Ford. However, seven-speed automatics are becoming available in some high-performance production luxury cars (found in Mercedes 7G gearbox, Infiniti), as are eight-speed autos in models from 2006 introduced by Aisin Seiki Co. in Lexus, ZF and Hyundai Motor Company. From 2013 are available nine speeds transmissions produced by ZF and Mercedes 9G.

Overdrive ('D', 'OD', or a boxed [D] or the absence of an illuminated 'O/D OFF'):

This mode is used in some transmissions to allow early computer-controlled transmissions to engage the automatic overdrive. In these transmissions, Drive (D) locks the automatic overdrive off, but is identical otherwise. OD (Overdrive) in these cars is engaged under steady speeds or low acceleration at approximately 35–45 mph (56–72 km/h). Under hard acceleration or below 35–45 mph (56–72 km/h), the transmission will automatically downshift. Other vehicles with this selector (for example light trucks) will not only disable up-shift to the overdrive gear, but keep the remaining available gears continuously engaged to the engine for use of compression braking. Drivers should verify the behaviour of this switch and consider the benefits of reduced friction brake use when city driving where speeds typically do not necessitate the overdrive gear.

Most automatic transmissions include some means of forcing a downshift (Throttle kickdown) into the lowest possible gear ratio if the throttle pedal is fully depressed. In many older designs, kickdown is accomplished by mechanically actuating a valve inside the transmission. Most modern designs use a solenoid-operated valve that is triggered by a switch on the throttle linkage or by the engine control unit (ECU) in response to an abrupt increase in engine power.

Mode selection allows the driver to choose between preset shifting programs. For example, Economy mode saves fuel by upshifting at lower engine speeds, while Sport mode (aka "Power" or "Performance") delays upshifting for maximum acceleration. Some transmission units also have Winter mode, where higher gear ratios are chosen to keep revs as low as possible while on slippery surfaces. The modes also change how the computer responds to throttle input.

Conventionally, automatic transmissions have selector positions that allow the driver to limit the maximum ratio that the transmission may engage. On older transmissions, this was accomplished by a mechanical lockout in the transmission valve body preventing an upshift until the lockout was disengaged; on computer-controlled transmissions, the same effect is accomplished by firmware. The transmission can still upshift and downshift automatically between the remaining ratios: for example, in the 3 range, a transmission could shift from first to second to third, but not into fourth or higher ratios. Some transmissions will still upshift automatically into the higher ratio if the engine reaches its maximum permissible speed in the selected range.

Third (3):

This mode limits the transmission to the first three gear ratios, or sometimes locks the transmission in third gear. This can be used to climb or going down hill. Some vehicles will automatically shift up out of third gear in this mode if a certain revolutions per minute

(RPM) range is reached in order to prevent engine damage. This gear is also recommended while towing a trailer.

Second (2 or S):

This mode limits the transmission to the first two gear ratios, or locks the transmission in second gear on Ford, Kia, and Honda models. This can be used to drive in adverse conditions such as snow and ice, as well as climbing or going down hills in winter. It is usually recommended to use second gear for starting on snow and ice, and use of this position enables this with an automatic transmission. Some vehicles will automatically shift up out of second gear in this mode if a certain RPM range is reached in order to prevent engine damage.

Although traditionally considered second gear, there are other names used. Chrysler models with a three-speed automatic since the late 1980s have called this gear 3 while using the traditional names for Drive and Low. Oldsmobile has called second gear as the 'Super' range — which was first used on their 4-speed Hydramatic transmissions, although the use of this term continued until the early 1980s when GM's Turbo Hydramatic automatic transmissions were standardized by all of their divisions years after the 4-speed Hydramatic was discontinued.

Some automatics, particularly those fitted to larger capacity or high torque engines, either when "2" is manually selected, or by engaging a winter mode, will start off in second gear instead of first, and then not shift into a higher gear until returned to "D." Also note that as with most American automatic transmissions, selecting "2" using the selection lever will not tell the transmission to be in only 2nd gear; rather, it will simply limit the transmission to 2nd gear after prolonging the duration of 1st gear through higher speeds than normal operation. The 2000–2002 Lincoln LS V8 (the five-speed automatic without manumatic capabilities, as opposed to the optional sport package w/ manu-matic 5-speed) started in 2nd gear during most starts both in winter and other seasons by selecting the "D5" transmission selection notch in the shiftgate (for fuel savings), whereas "D4" would always start in 1st gear. This is done to reduce torque multiplication when proceeding forward from a standstill in conditions where traction was limited — on snow- or ice-covered roads, for example.

First (1 or L [Low]):

This mode locks the transmission in first gear only. In older vehicles, it will not change to any other gear range. Some vehicles will automatically shift up out of first gear in this mode if a certain RPM range is reached in order to prevent engine damage. This, like second, can be used during the winter season, for towing, or for downhill driving to increase the engine braking effect. The "Austin Mini" automatic transmission is different in this respect - This mode locks the transmission in first gear, but the gearbox has a freewheel on the overrun. Closing the throttle after acceleration results in the vehicle continuing at the same speed and only slowing down due to friction and wind resistance. During this time, the engine RPM will drop back to idle until the throttle is pressed again. What this means is that in "First", engine braking is not available and "2" is the lowest gear that should be used whilst descending hills.

One type of manumatic shifting system available on automatic transmissions are paddle shifters. The paddle depicted here is the upshift paddle in a 2013 Honda Accord, with the driver's hand on it. Manumatics and paddle shifters may control any type of automatic transmission, including the continuously variable transmission in the Accord.

Manual Controls

Some transmissions have a mode in which the driver has full control of ratios change (either by moving the selector, or through the use of buttons or paddles), completely overriding the automated function of the hydraulic controller. Such control is particularly useful in cornering, to avoid unwanted upshifts or downshifts that could compromise the vehicle's balance or traction. "Manumatic" shifters, first popularized by Porsche in the 1990s under the trade name Tiptronic, have become a popular option on sports cars and other performance vehicles. With the near-universal prevalence of electronically controlled transmissions, they are comparatively simple and inexpensive, requiring only software changes, and the provision of the actual manual controls for the driver. The amount of true manual control provided is highly variable: some systems will override the driver's selections under certain conditions, generally in the interest of preventing engine damage. Since these gearboxes also have a throttle kickdown switch, it is impossible to fully exploit the engine power at low to medium engine speeds.

Manufacturer-specific Modes

As well as the above modes there are other modes, dependent on the manufacturer and model. Some examples include:

D5:

> In Hondas and Acuras equipped with five-speed automatic transmissions, this mode is used commonly for highway use (as stated in the manual), and uses all five forward gear ratios.

D4:

> This mode is also found in Honda and Acura four or five-speed automatics, and only uses

the first four gear ratios. According to the manual, it is used for stop-and-go traffic, such as city driving.

D3 or 3:

This mode is found in Honda, Acura, Volkswagen and Pontiac four-speed automatics and only uses the first three gear ratios. According to the manual, it is used for stop-and-go traffic, such as city driving.

D2 and D1:

These modes are found on older Ford transmissions (C6, etc.). In D1, all three gears are used, whereas in D2 the car starts in second gear and upshifts to third.

S or Sport:

This is commonly described as Sport mode. It operates in an identical manner as "D" mode, except that the upshifts change much higher up the engine's rev range. This has the effect on maximising all the available engine output, and therefore enhances the performance of the vehicle, particularly during acceleration. This mode will also downchange much higher up the rev range compared to "D" mode, maximising the effects of engine braking. This mode will have a detrimental effect on fuel economy. Hyundai has a Norm/Power switch next to the gearshift for this purpose on the Tiburon.

Some early GMs equipped with HYDRA-MATIC transmissions used (S) to indicate Second gear, being the same as the 2 position on a Chrysler, shifting between only first and second gears. This would have been recommended for use on steep grades, or slippery roads like dirt, or ice, and limited to speeds under 40 mph. (L) was used in some early GMs to indicate (L)ow gear, being the same as the 2 position on a Chrysler, locking the transmission into first gear. This would have been recommended for use on steep grades, or slippery roads like dirt, or ice, and limited to speeds under 15 mph.

+ –, and M:

This is for the Manual mode selection of gears in certain automatics, such as Porsche and Honda's Tiptronic and BMW and Kia's Steptronic. The M feature can also be found in vehicles such as the Dodge Magnum and Journey; Pontiac G6; Mazda3, Mazda6, and CX-7; Toyota Camry, Corolla, Fortuner, Previa and Innova; Kia Forte (K3/Cerato), Optima (K5), Cadenza (K7) and K9 (Quoris). Mitsubishi montero sport / pajero sport and some Audi models (Audi TT) do not have the M, and instead have the + and -, which is separated from the rest of the shift modes; the same is true for some Peugeot products like the Peugeot 206. Meanwhile, the driver can shift up and down at will by toggling the (console mounted) shift lever similar to a semi-automatic transmission. This mode may be engaged either through a selector/position or by actually changing the gears (e.g., tipping the gear-down paddles mounted near the driver's fingers on the steering wheel).

Winter (W):

In some Volvo, Mercedes-Benz, BMW and General Motors Europe models, a winter mode can be engaged so that second gear is selected instead of first when pulling away from stationary, to reduce the likelihood of loss of traction due to wheel spin on snow or ice. On GM cars, this was D2 in the 1950s, and is Second Gear Start after 1990. On Ford, Kia, and Honda automatics, this feature can be accessed by moving the gear selector to 2 to start, then taking your foot off the accelerator while selecting D once the car is moving.

Brake (B):

A mode selectable on some Toyota models, as well as electric cars from several manufacturers. It can be used to decelerate, or maintain speed going downhill, without using the conventional brakes. In non-hybrid cars, B mode selects a lower gear to increase engine braking. GM called this "HR" ("hill retarder") and "GR" ("grade retarder") in the 1950s. In hybrids such as the Toyota Prius, which have a fixed gear ratio, B mode slows the car in part by increasing engine air intake, which enhances engine braking. In electric cars such as the Nissan Leaf and Mitsubishi i-MiEV, B mode increases the level of regenerative braking when the accelerator pedal is released.

Some automatic transmissions modified or designed specifically for drag racing may also incorporate a transbrake as part of a manual valve body. Activated by electrical solenoid control, a transbrake simultaneously engages the first and reverse gears, locking the transmission and preventing the input shaft from turning. This allows the driver of the car to raise the engine RPM against the resistance of the torque converter, then launch the car by simply releasing the transbrake switch.

Comparison with Manual Transmission

Most cars sold in North America since the 1950s have been available with an automatic transmission based on the fact that the three major American car manufacturers had started using automatics. Conversely, in Europe a manual gearbox is standard, with 20% of drivers opting for an automatic transmission. In some Asian markets and in Australia, automatic transmissions have become very popular since the 1980s.

Vehicles equipped with automatic transmissions are not as complex to drive. Consequently, in some jurisdictions, drivers who have passed their driving test in a vehicle with an automatic transmission will not be licensed to drive a manual transmission vehicle. Conversely, a manual license will allow the driver to drive vehicles with either transmission. Countries in which such driving license restrictions are applied include some states in Australia, Austria, Belgium, Belize, China, Croatia, Denmark, Dominican Republic, Estonia, Finland, France, Germany, Hong Kong, Hungary, India, Indonesia, Ireland, Israel, Japan, Latvia, Lebanon, Lithuania, Macau, Mauritius, the Netherlands, New Zealand (restricted licence only), Norway, Philippines, Poland, Portugal, Qatar, Romania, Russia (as of April 2014), Saudi Arabia (as of March 2012), Singapore, Slovenia, South Africa, South Korea, Spain, Sweden, Switzerland, Taiwan, Trinidad and Tobago, United Arab Emirates and the United Kingdom.

A conventional manual transmission is frequently the base equipment in a car, with the option being an automated transmission such as a conventional automatic, semi-automatic, or CVT.

Effects on Vehicle Control

Cornering

Unexpected gear changes can affect the attitude of a vehicle in marginal conditions.

Maintaining Constant Speed

Torque converters and CVT transmissions make changes in vehicle speed less apparent by the engine noise, as they decouple the engine speed from vehicle speed.

Lockup torque converters that engage and disengage at certain speeds can make these speeds unstable — the transmission wastes less power above the speed at which the torque converter locks up, usually causing more power to the wheels for the same throttle input.

Controlling Wheelspin

Torque converters respond quickly to loss of traction (torque) by an increased speed of the driving wheels for the same engine speed. Thus, under most conditions, where the static friction is higher than the kinetic friction, the engine speed must be brought down to counteract wheelspin when it has occurred, requiring a stronger or quicker throttle reduction by the driver than with a manual transmission, making wheelspin harder to control. This is most apparent in driving conditions with much higher static friction than kinetic, such as packed hard snow (that turns to ice by friction work), or snow on top of ice.

Climbing Steep Slippery Slopes

In situations where a driver with a manual transmission can't afford a gearshift, in fear of losing too much speed to reach a hilltop, automatic transmissions are at a great advantage — whereas the manual driver depends on finding a gear that is not too low to enter the bottom of the hill at the necessary speed, but not too high to stall the engine at the top of the hill, sometimes an impossible task, this is a non-issue with automatic transmissions, not just because gearshifts are quick, but they typically maintain some power on the driving wheels during the gearshift.

Energy Efficiency:

Earlier hydraulic automatic transmissions were almost always less energy efficient than manual transmissions due mainly to viscous and pumping losses, both in the torque converter and the hydraulic actuators. 21% is the loss on a 3 speed Chrysler Torqueflite compared to a modern GM 6L80 automatic. A relatively small amount of energy is required to pressurize the hydraulic control system, which uses fluid pressure to determine the correct shifting patterns and operate the various automatic clutch mechanisms. However, with technological developments some modern Continuously variable transmission are more fuel efficient than their manual counterparts and modern 8 speed automatics are within 5% as efficient as a manual gearbox.

Manual transmissions use a mechanical clutch to transmit torque, rather than a torque converter, thus avoiding the primary source of loss in an automatic transmission. Manual transmissions also avoid the power requirement of the hydraulic control system, by relying on the human muscle power of the vehicle operator to disengage the clutch and actuate the gear levers, and the mental power of the operator to make appropriate gear ratio selections. Thus the manual transmission requires very little engine power to function, with the main power consumption due to drag from the gear train being immersed in the lubricating oil of the gearbox.

The on-road acceleration of an automatic transmission can occasionally exceed that of an otherwise identical vehicle equipped with a manual transmission in turbocharged diesel applications. Turbo-boost is normally lost between gear changes in a manual whereas in an automatic the accelerator pedal can remain fully depressed. This however is still largely dependent upon the number and optimal spacing of gear ratios for each unit, and whether or not the elimination of spooldown/ accelerator lift off represent a significant enough gain to counter the slightly higher power consumption of the automatic transmission itself.

Automatic Transmission Models

Some of the best known automatic transmission families include:

- General Motors — Dynaflow, Powerglide, Turboglide, "Turbo-Hydramatic" TH350, TH400 and 700R4, 4L60-E, 4L80-E, Holden Trimatic.
- Ford: Cruise-O-Matic, C4, CD4E, C6, AOD/AODE, E4OD, ATX, AXOD/AX4S/AX4N.
- Cummins 68 RFE (fitted to the Ram diesel segment).
- Chrysler: TorqueFlite 727 and 904, A500, A518, 45RFE, 545RFE.
- BorgWarner (later Aisin AW).
- ZF Friedrichshafen automatic transmissions.
- Mercedes-Benz transmissions.
- Allison Transmission.
- Voith Voith Turbo.
- Aisin AW; Aisin AW is a Japanese automotive parts supplier, known for its automatic transmissions and navigation systems.
- Honda.
- Nissan/Jatco.
- Volkswagen Group — 01M.
- Drivetrain Systems International (DSI) — M93, M97 and M74 4-speeds, M78 and M79 6-speeds.
- Hyundai Hyundai Powertech — 4F12, 4F16, 4F23 4-Speeds, 5F25, 5F16, 5F23 5-Speeds, 6F17, 6F26, 6F40 6-Speeds, 8R40, 8R50 8-Speeds, Mini Cooper — Automatic or manual transmission, all models.

Automatic transmission families are usually based on Ravigneaux, Lepelletier, or Simpson planetary gearsets. Each uses some arrangement of one or two central sun gears, and a ring gear, with differing arrangements of planet gears that surround the sun and mesh with the ring. An exception to this is the Hondamatic line from Honda, which uses sliding gears on parallel axes like a manual transmission without any planetary gearsets. Although the Honda is quite different from all other automatics, it is also quite different from an automated manual transmission (AMT).

Many of the above AMTs exist in modified states, which were created by racing enthusiasts and their mechanics by systematically re-engineering the transmission to achieve higher levels of performance. These are known as "performance transmissions". Example of manufacturers of high performance transmissions are General Motors and Ford.

Manufacturing Engineering

Manufacturing engineering is a discipline of engineering dealing with various manufacturing sciences and practices including the research, design and development of systems, processes, machines, tools, and equipment. The manufacturing engineer's primary focus is to turn raw materials into a new or updated product in the most economic, efficient, and effective way possible.

Overview

This field also deals with the integration of different facilities and systems for producing quality products (with optimal expenditure) by applying the principles of physics and the results of manufacturing systems studies, such as the following:

• Craft or Guild	• Computer integrated manufacturing	• Agile manufacturing
• Putting-out system	• Computer-aided technologies in manufacturing	• Rapid manufacturing
• British factory system	• Just in time manufacturing	• Prefabrication
• American system of manufacturing	• Lean manufacturing	• Ownership
• Soviet collectivism in manufacturing	• Flexible manufacturing	• Fabrication
• Mass production	• Mass customization	• Publication

A set of six-axis robots used for welding.

Manufacturing engineers develop and create physical artifacts, production processes, and technology. It is a very broad area which includes the design and development of products. Manufacturing engineering is considered to be a subdiscipline of industrial engineering/systems engineering and has very strong overlaps with mechanical engineering. Manufacturing engineers' success or failure directly impacts the advancement of technology and the spread of innovation. This field of manufacturing engineering emerged from tool and die discipline in the early 20th century. It expanded greatly from the 1960s when industrialized countries introduced factories with:

1. Numerical control machine tools and automated systems of production.

2. Advanced statistical methods of quality control: These factories were pioneered by the American electrical engineer William Edwards Deming, who was initially ignored by his home country. The same methods of quality control later turned Japanese factories into world leaders in cost-effectiveness and production quality.

3. Industrial robots on the factory floor, introduced in the late 1970s: These computer-controlled welding arms and grippers could perform simple tasks such as attaching a car door quickly and flawlessly 24 hours a day. This cut costs and improved production speed.

History

The history of manufacturing engineering can be traced to factories in the mid 19th century USA and 18th century UK. Although large home production sites and workshops were established in China, ancient Rome and the Middle East, the Venice Arsenal provides one of the first examples of a factory in the modern sense of the word. Founded in 1104 in the Republic of Venice several hundred years before the Industrial Revolution, this factory mass-produced ships on assembly lines using manufactured parts. The Venice Arsenal apparently produced nearly one ship every day and, at its height, employed 16,000 people.

Many historians regard Matthew Boulton's Soho Manufactory (established in 1761 in Birmingham) as the first modern factory. Similar claims can be made for John Lombe's silk mill in Derby (1721), or Richard Arkwright's Cromford Mill (1771). The Cromford Mill was purpose-built to accommodate the equipment it held and to take the material through the various manufacturing processes.

Ford assembly line, 1913.

One historian, Murno Gladst, contends that the first factory was in Potosí. The Potosi factory took advantage of the abundant silver that was mined nearby and processed silver ingot slugs into coins.

British colonies in the 19th century built factories simply as buildings where a large number of workers gathered to perform hand labor, usually in textile production. This proved more efficient for the administration and distribution of materials to individual workers than earlier methods of manufacturing, such as cottage industries or the putting-out system.

Cotton mills used inventions such as the steam engine and the power loom to pioneer the industrial factories of the 19th century, where precision machine tools and replaceable parts allowed greater efficiency and less waste. This experience formed the basis for the later studies of manufacturing engineering. Between 1820 and 1850, non-mechanized factories supplanted traditional artisan shops as the predominant form of manufacturing institution.

Henry Ford further revolutionized the factory concept and thus manufacturing engineering in the early 20th century with the innovation of mass production. Highly specialized workers situated alongside a series of rolling ramps would build up a product such as (in Ford's case) an automobile. This concept dramatically decreased production costs for virtually all manufactured goods and brought about the age of consumerism.

Modern Developments

Modern manufacturing engineering studies include all intermediate processes required for the production and integration of a product's components.

Some industries, such as semiconductor and steel manufacturers use the term "fabrication" for these processes.

KUKA industrial robots being used at a bakery for food production.

Automation is used in different processes of manufacturing such as machining and welding. Automated manufacturing refers to the application of automation to produce goods in a factory. The main advantages of automated manufacturing for the manufacturing process are realized with effective implementation of automation and include: higher consistency and quality, reduction of lead times, simplification of production, reduced handling, improved work flow, and improved worker morale.

Robotics is the application of mechatronics and automation to create robots, which are often used in manufacturing to perform tasks that are dangerous, unpleasant, or repetitive. These robots may be of any shape and size, but all are preprogrammed and interact physically with the world. To create a robot, an engineer typically employs kinematics (to determine the robot's range of motion) and mechanics (to determine the stresses within the robot). Robots are used extensively in manufacturing engineering.

Robots allow businesses to save money on labor, perform tasks that are either too dangerous or too precise for humans to perform economically, and to ensure better quality. Many companies employ assembly lines of robots, and some factories are so robotized that they can run by themselves. Outside the factory, robots have been employed in bomb disposal, space exploration, and many other fields. Robots are also sold for various residential applications.

Education

Certification Programs

Manufacturing engineers possess a bachelor's degree in engineering with a major in manufacturing engineering. The length of study for such a degree is usually four to five years followed by five more years of professional practice to qualify as a professional engineer. Working as a manufacturing engineering technologist involves a more applications-oriented qualification path.

Academic degrees for manufacturing engineers are usually the Bachelor of Engineering, [BE] or [BEng], and the Bachelor of Science, [BS] or [BSc]. For manufacturing technologists the required degrees are Bachelor of Technology [B.TECH] or Bachelor of Applied Science [BASc] in Manufacturing, depending upon the university. Master's degrees in engineering manufacturing include Master of Engineering [ME] or [MEng] in Manufacturing, Master of Science [M.Sc] in Manufacturing Management, Master of Science [M.Sc] in Industrial and Production Management, and Master of Science [M.Sc] as well as Master of Engineering [ME] in Design, which is a subdiscipline of manufacturing. Doctoral [PhD] or [DEng] level courses in manufacturing are also available depending on the university.

The undergraduate degree curriculum generally includes courses in physics, mathematics, computer science, project management, and specific topics in mechanical and manufacturing engineering. Initially such topics cover most, if not all, of the subdisciplines of manufacturing engineering. Students then choose to specialize in one or more subdisciplines towards the end of their degree work.

Syllabus

The foundational curriculum for a bachelor's degree in manufacturing engineering is very similar to that for mechanical engineering, and includes:

- Statics and dynamics.
- Strength of materials and solid mechanics.
- Instrumentation and measurement.

- Applied thermodynamics, heat transfer, energy conversion, and HVAC.
- Fluid mechanics and fluid dynamics.
- Mechanism design (including kinematics and dynamics).
- Manufacturing technology or processes.
- Hydraulics and pneumatics.
- Mathematics - in particular, calculus, differential equations, statistics, and linear algebra.
- Engineering design and graphics.
- Circuit Analysis.
- Lean manufacturing.
- Mechatronics and control theory.
- Automation and reverse engineering.
- Quality assurance and control.
- Material science.
- Drafting, CAD (including solid modeling), and CAM, etc.

A bachelor's degree in these two areas will typically differ only by a few specialized classes, although the mechanical engineering degree requires more mathematics expertise.

Manufacturing Engineering Certification

Certification and licensure:

In some countries, "professional engineer" is the term for registered or licensed engineers who are permitted to offer their professional services directly to the public. Professional Engineer, abbreviated (PE - USA) or (PEng - Canada), is the designation for licensure in North America. In order to qualify for this license, a candidate needs a bachelor's degree from an ABET recognized university in the USA, a passing score on a state examination, and four years of work experience usually gained via a structured internship. In the USA, more recent graduates have the option of dividing this licensure process into two segments. The Fundamentals of Engineering (FE) exam is often taken immediately after graduation and the Principles and Practice of Engineering exam is taken after four years of working in a chosen engineering field.

Society of Manufacturing Engineers (SME) certifications (USA):

The SME administers qualifications specifically for the manufacturing industry. These are not degree level qualifications and are not recognized at the professional engineering level. The following discussion deals with qualifications in the USA only. Qualified candidates for the Certified Manufacturing Technologist Certificate (CMfgT) must pass a three-hour, 130-question multiple-choice exam. The exam covers math, manufacturing processes, manufacturing management, automation, and related subjects. Additionally, a candidate must have at least four years of combined education and manufacturing-related work experience.

Certified Manufacturing Engineer (CMfgE) is an engineering qualification administered by the Society of Manufacturing Engineers, Dearborn, Michigan, USA. Candidates qualifying for a Certified Manufacturing Engineer credential must pass a four-hour, 180 question multiple-choice exam which covers more in-depth topics than does the CMfgT exam. CMfgE candidates must also have eight years of combined education and manufacturing-related work experience, with a minimum of four years of work experience.

Certified Engineering Manager (CEM). The Certified Engineering Manager Certificate is also designed for engineers with eight years of combined education and manufacturing experience. The test is four hours long and has 160 multiple-choice questions. The CEM certification exam covers business processes, teamwork, responsibility, and other management-related categories.

Modern Tools

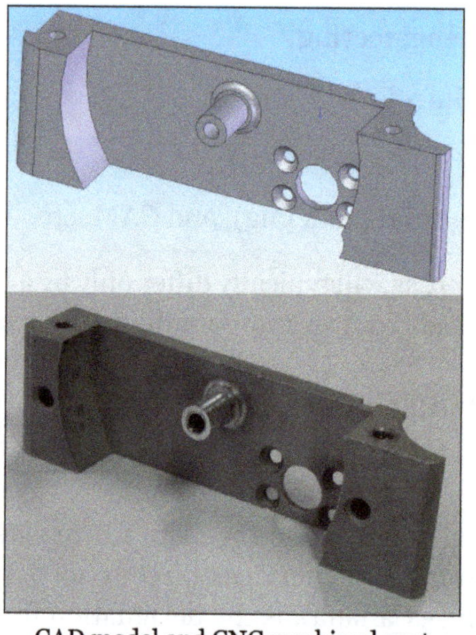

CAD model and CNC machined part.

Many manufacturing companies, especially those in industrialized nations, have begun to incorporate computer-aided engineering (CAE) programs into their existing design and analysis processes, including 2D and 3D solid modeling computer-aided design (CAD). This method has many benefits, including easier and more exhaustive visualization of products, the ability to create virtual assemblies of parts, and ease of use in designing mating interfaces and tolerances.

Other CAE programs commonly used by product manufacturers include product life cycle management (PLM) tools and analysis tools used to perform complex simulations. Analysis tools may be used to predict product response to expected loads, including fatigue life and manufacturability. These tools include finite element analysis (FEA), computational fluid dynamics (CFD), and computer-aided manufacturing (CAM).

Using CAE programs, a mechanical design team can quickly and cheaply iterate the design process to develop a product that better meets cost, performance, and other constraints. No physical prototype need be created until the design nears completion, allowing hundreds or thousands of designs to be evaluated, instead of relatively few. In addition, CAE analysis programs can model complicated physical phenomena which cannot be solved by hand, such as viscoelasticity, complex contact between mating parts, or non-Newtonian flows.

Just as manufacturing engineering is linked with other disciplines, such as mechatronics, multidisciplinary design optimization (MDO) is also being used with other CAE programs to automate and improve the iterative design process. MDO tools wrap around existing CAE processes, allowing product evaluation to continue even after the analyst goes home for the day. They also utilize sophisticated optimization algorithms to more intelligently explore possible designs, often finding better, innovative solutions to difficult multidisciplinary design problems.

Subdisciplines

Mechanics

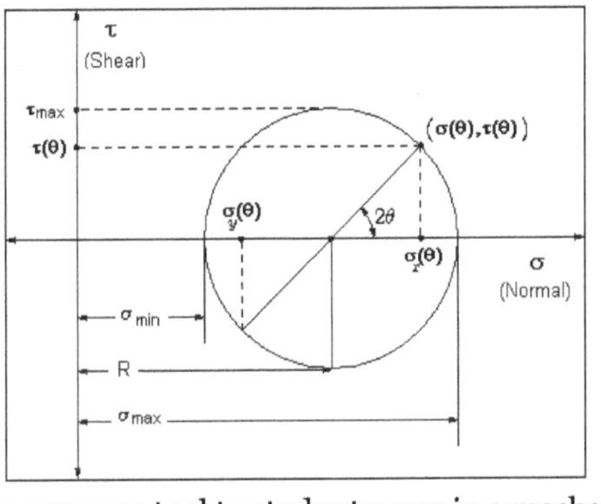

Mohr's circle, a common tool to study stresses in a mechanical element.

Mechanics, in the most general sense, is the study of forces and their effects on matter. Typically, engineering mechanics is used to analyze and predict the acceleration and deformation (both elastic and plastic) of objects under known forces (also called loads) or stresses. Subdisciplines of mechanics include:

- Statics, the study of non-moving bodies under known loads.
- Dynamics (or kinetics), the study of how forces affect moving bodies.
- Mechanics of materials, the study of how different materials deform under various types of stress.
- Fluid mechanics, the study of how fluids react to forces.
- Continuum mechanics, a method of applying mechanics that assumes that objects are continuous (rather than discrete).

If the engineering project were to design a vehicle, statics might be employed to design the frame of the vehicle in order to evaluate where the stresses will be most intense. Dynamics might be used when designing the car's engine to evaluate the forces in the pistons and cams as the engine cycles. Mechanics of materials might be used to choose appropriate materials for the manufacture of the frame and engine. Fluid mechanics might be used to design a ventilation system for the vehicle or to design the intake system for the engine.

Kinematics

Kinematics is the study of the motion of bodies (objects) and systems (groups of objects), while ignoring the forces that cause the motion. The movement of a crane and the oscillations of a piston in an engine are both simple kinematic systems. The crane is a type of open kinematic chain, while the piston is part of a closed four-bar linkage. Engineers typically use kinematics in the design and analysis of mechanisms. Kinematics can be used to find the possible range of motion for a given mechanism, or, working in reverse, can be used to design a mechanism that has a desired range of motion.

Drafting

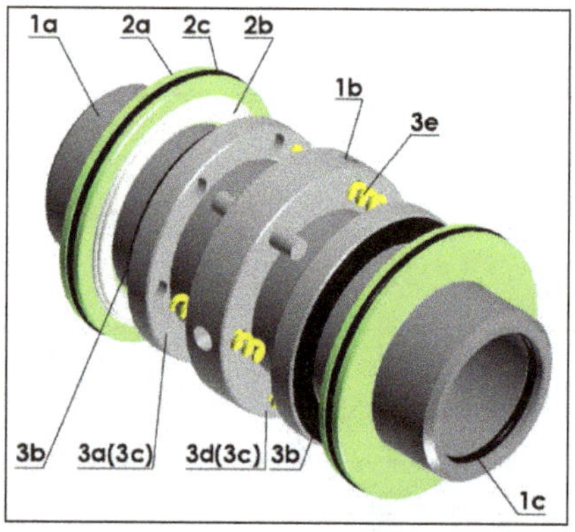

A CAD model of a mechanical double seal.

Drafting or technical drawing is the means by which manufacturers create instructions for manufacturing parts. A technical drawing can be a computer model or hand-drawn schematic showing all the dimensions necessary to manufacture a part, as well as assembly notes, a list of required materials, and other pertinent information. A U.S engineer or skilled worker who creates technical drawings may be referred to as a drafter or draftsman. Drafting has historically been a two-dimensional process, but computer-aided design (CAD) programs now allow the designer to create in three dimensions.

Instructions for manufacturing a part must be fed to the necessary machinery, either manually, through programmed instructions, or through the use of a computer-aided manufacturing (CAM) or combined CAD/CAM program. Optionally, an engineer may also manually manufacture a part using the technical drawings, but this is becoming an increasing rarity with the advent of computer numerically controlled (CNC) manufacturing. Engineers primarily manufacture parts manually in the areas of applied spray coatings, finishes, and other processes that cannot economically or practically be done by a machine.

Drafting is used in nearly every subdiscipline of mechanical and manufacturing engineering, and by many other branches of engineering and architecture. Three-dimensional models created using CAD software are also commonly used in finite element analysis (FEA) and computational fluid dynamics (CFD).

Machine Tools and Metal Fabrication

Machine tools employ some sort of tool that does the cutting or shaping. All machine tools have some means of constraining the workpiece and provide a guided movement of the parts of the machine. Metal fabrication is the building of metal structures by cutting, bending, and assembling processes.

Computer Integrated Manufacturing

Computer-integrated manufacturing (CIM) is the manufacturing approach of using computers to control the entire production process. Computer-integrated manufacturing is used in automotive, aviation, space, and ship building industries.

Mechatronics

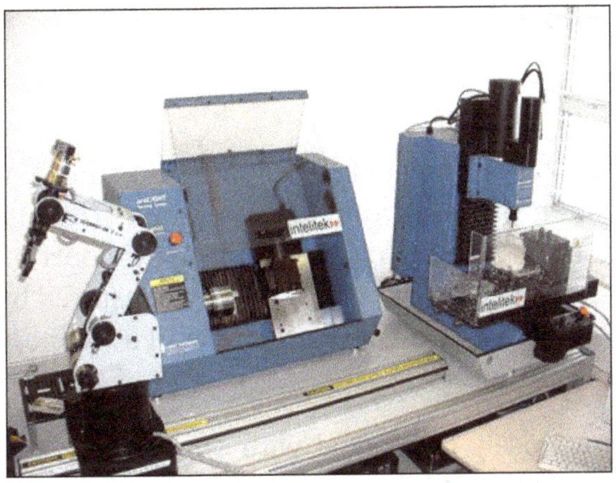

Training FMS with learning robot SCORBOT-ER 4u, workbench CNC mill and CNC lathe.

Mechatronics is an engineering discipline that deals with the convergence of electrical, mechanical and manufacturing systems. Such combined systems are known as electromechanical systems and are widespread. Examples include automated manufacturing systems, heating, ventilation and air-conditioning systems, and various aircraft and automobile subsystems.

The term mechatronics is typically used to refer to macroscopic systems, but futurists have predicted the emergence of very small electromechanical devices. Already such small devices, known as Microelectromechanical systems (MEMS), are used in automobiles to initiate the deployment of airbags, in digital projectors to create sharper images, and in inkjet printers to create nozzles for high-definition printing. In the future it is hoped that such devices will be used in tiny implantable medical devices and to improve optical communication.

Textile Engineering

Textile engineering courses deal with the application of scientific and engineering principles to the design and control of all aspects of fiber, textile, and apparel processes, products, and machinery. These include natural and man-made materials, interaction of materials with machines, safety and health, energy conservation, and waste and pollution control. Additionally, students are given experience in plant design and layout, machine and wet process design and improvement, and designing and creating textile products. Throughout the textile engineering curriculum, students take classes from other engineering and disciplines including: mechanical, chemical, materials and industrial engineering.

Advanced Composite Materials

Advanced composite materials (engineering) (ACMs) are also known as advanced polymer matrix composites. These are generally characterized or determined by unusually high strength fibres with unusually high stiffness, or modulus of elasticity characteristics, compared to other materials, while bound together by weaker matrices. Advanced composite materials have broad, proven applications, in the aircraft, aerospace, and sports equipment sectors. Even more specifically ACMs are very attractive for aircraft and aerospace structural parts. Manufacturing ACMs is a multibillion-dollar industry worldwide. Composite products range from skateboards to components of the space shuttle. The industry can be generally divided into two basic segments, industrial composites and advanced composites.

Employment

Manufacturing engineering is just one facet of the engineering industry. Manufacturing engineers enjoy improving the production process from start to finish. They have the ability to keep the whole production process in mind as they focus on a particular portion of the process. Successful students in manufacturing engineering degree programs are inspired by the notion of starting with a natural resource, such as a block of wood, and ending with a usable, valuable product, such as a desk, produced efficiently and economically.

Manufacturing engineers are closely connected with engineering and industrial design efforts. Examples of major companies that employ manufacturing engineers in the United States include General Motors Corporation, Ford Motor Company, Chrysler, Boeing, Gates Corporation and Pfizer. Examples in Europe include Airbus, Daimler, BMW, Fiat, Navistar International, and Michelin Tyre.

Industries where manufacturing engineers are generally employed include:

- Aerospace industry
- Automotive industry
- Chemical industry
- Computer industry
- Food processing industry
- Garment industry
- Pharmaceutical industry
- Pulp and paper industry
- Toy industry

Frontiers of Research

Flexible Manufacturing Systems

A typical FMS system.

A flexible manufacturing system (FMS) is a manufacturing system in which there is some amount of flexibility that allows the system to react to changes, whether predicted or unpredicted. This flexibility is generally considered to fall into two categories, both of which have numerous subcategories. The first category, machine flexibility, covers the system's ability to be changed to produce new product types and the ability to change the order of operations executed on a part. The second category, called routing flexibility, consists of the ability to use multiple machines to perform the same operation on a part, as well as the system's ability to absorb large-scale changes, such as in volume, capacity, or capability.

Most FMS systems comprise three main systems. The work machines, which are often automated CNC machines, are connected by a material handling system to optimize parts flow, and to a central control computer, which controls material movements and machine flow. The main advantages of an FMS is its high flexibility in managing manufacturing resources like time and effort in order to manufacture a new product. The best application of an FMS is found in the production of small sets of products from a mass production.

Computer Integrated Manufacturing

Computer-integrated manufacturing (CIM) in engineering is a method of manufacturing in which the entire production process is controlled by computer. Traditionally separated process methods are joined through a computer by CIM. This integration allows the processes to exchange information and to initiate actions. Through this integration, manufacturing can be faster and less error-prone, although the main advantage is the ability to create automated manufacturing processes. Typically CIM relies on closed-loop control processes based on real-time input from sensors. It is also known as flexible design and manufacturing.

Friction Stir Welding

Friction stir welding was discovered in 1991 by The Welding Institute (TWI). This innovative steady state (non-fusion) welding technique joins previously un-weldable materials, including several aluminum alloys. It may play an important role in the future construction of airplanes, potentially replacing rivets. Current uses of this technology to date include: welding the seams of the aluminum main space shuttle external tank, the Orion Crew Vehicle test article, Boeing Delta II and Delta IV Expendable Launch Vehicles and the SpaceX Falcon 1 rocket; armor plating for amphibious assault ships; and welding the wings and fuselage panels of the new Eclipse 500 aircraft from Eclipse Aviation, among an increasingly growing range of uses.

Close-up view of a friction stir weld tack tool.

Other areas of research are Product Design, MEMS (Micro-Electro-Mechanical Systems), Lean Manufacturing, Intelligent Manufacturing Systems, Green Manufacturing, Precision Engineering, Smart Materials, etc.

g-force

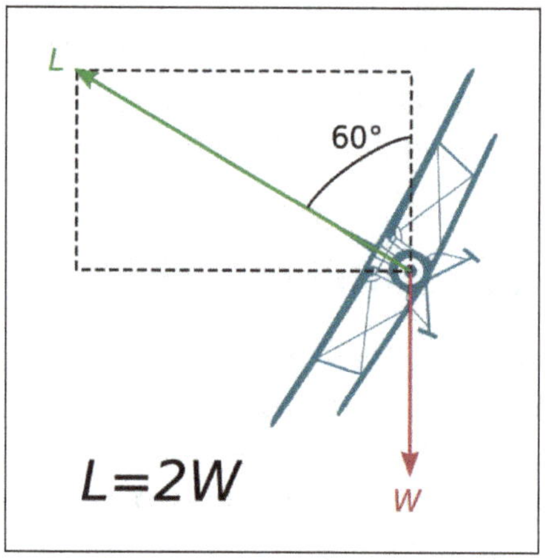

In straight and level flight, lift (L) equals weight (W). In a banked turn of 60°, lift equals double the weight (L=2W). The pilot experiences 2 g and a doubled weight. The steeper the bank, the greater the g-forces.

g-force (with g from gravitational) is a measurement of the type of acceleration that causes weight. Despite the name, it is incorrect to consider g-force a fundamental force, as "g-force" (lower case character) is a type of acceleration that can be measured with an accelerometer. Since g-force accelerations indirectly produce weight, any g-force can be described as a "weight per unit mass". When the g-force acceleration is produced by the surface of one object being pushed by the surface of another object, the reaction-force to this push produces an equal and opposite weight for every unit of an object's mass. The types of forces involved are transmitted through objects by interior mechanical stresses. The g-force acceleration (save for certain electromagnetic force influences) is the cause of an object's acceleration in relation to free-fall.

This top-fuel dragster can accelerate from zero to 160 kilometres per hour (99 mph) in 0.86 seconds. This is a horizontal acceleration of 5.3 g. Combined with the vertical g-force in the stationary case the Pythagorean theorem yields a g force of 5.4 g.

The g-force acceleration experienced by an object is due to the vector sum of all non-gravitational and non-electromagnetic forces acting on an object's freedom to move. In practice, as noted, these are surface-contact forces between objects. Such forces cause stresses and strains on objects, since they must be transmitted from an object surface. Because of these strains, large g-forces may be destructive.

Gravitation acting alone does not produce a g-force, even though g-forces are expressed in multiples of the acceleration of a standard gravity. Thus, the standard gravitational acceleration at the Earth's surface produces g-force only indirectly, as a result of resistance to it by mechanical forces. These mechanical forces actually produce the g-force acceleration on a mass. For example, the 1 g force on an object sitting on the Earth's surface is caused by mechanical force exerted in the upward direction by the ground, keeping the object from going into free-fall. The upward contact-force from the ground ensures that an object at rest on the Earth's surface is accelerating relative to the free-fall condition. (Free fall is the path that the object would follow when falling freely toward the Earth's center). Stress inside the object is ensured from the fact that the ground contact forces are transmitted only from the point of contact with the ground.

Objects allowed to free-fall in an inertial trajectory under the influence of gravitation only, feel no g-force acceleration, a condition known as zero-g (which means zero g-force). This is demonstrated by the "zero-g" conditions inside a freely falling elevator falling toward the Earth's center (in vacuum), or (to good approximation) conditions inside a spacecraft in Earth orbit. These are examples of coordinate acceleration (a change in velocity) without a sensation of weight. The experience of no g-force (zero-g), however it is produced, is synonymous with weightlessness.

In the absence of gravitational fields, or in directions at right angles to them, proper and coordinate accelerations are the same, and any coordinate acceleration must be produced by a corresponding g-force acceleration. An example here is a rocket in free space, in which simple changes in velocity are produced by the engines, and produce g-forces on the rocket and passengers.

Unit and Measurement

The unit of measure of acceleration in the International System of Units (SI) is m/s^2. However, to distinguish acceleration relative to free-fall from simple acceleration (rate of change of velocity), the unit g (or g) is often used. One g is the acceleration due to gravity at the Earth's surface and is the standard gravity (symbol: g_n), defined as 9.80665 metres per second squared, or equivalently 9.80665 newtons of force per kilogram of mass. Note that the unit definition does not vary with location—the g-force when standing on the moon is about 0.181 g.

The unit g is not one of the SI units, which uses "g" for gram. Also "g" should not be confused with "G", which is the standard symbol for the gravitational constant. This notation is commonly used in aviation, especially in acrobatic or combat military aviation, to describe the increased forces that must be overcome by pilots in order to remain conscious and not G-LOC (G-induced loss of consciousness). For example, it is often said an F-16 fighter jet is able to sustain up to 9 G's for a limited time.

Measurement of g-force is typically achieved using an accelerometer. In certain cases, g-forces may be measured using suitably calibrated scales. Specific force is another name that has been used for g-force.

Acceleration and Forces

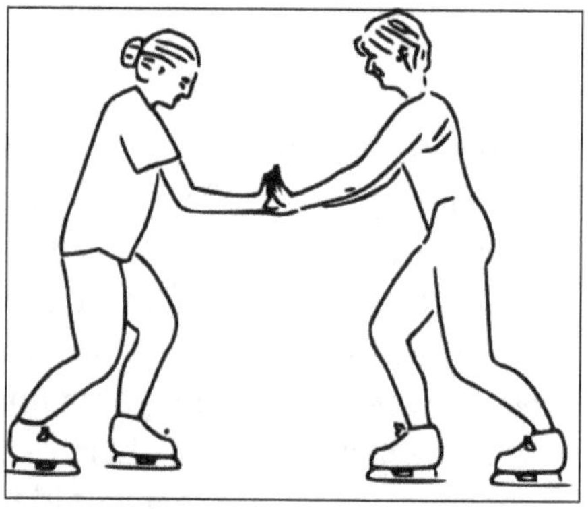

Newton's third law: law of reciprocal actions.

The term g-force is technically incorrect as it is a measure of acceleration, not force. While acceleration is a vector quantity, g-force accelerations ("g-forces" for short) are often expressed as a scalar, with positive g-forces pointing downward (indicating upward acceleration), and negative g-forces pointing upward. Thus, a g-force is a vector acceleration. It is an acceleration that must be produced by a mechanical force, and cannot be produced by simple gravitation. Objects acted upon only by gravitation, experience (or "feel") no g-force, and are weightless.

G-forces, when multiplied by a mass upon which they act, are associated with a certain type of mechanical force in the correct sense of the term force, and this force produces compressive stress and tensile stress. Such forces result in the operational sensation of weight, but the equation carries a sign change due to the definition of positive weight in the direction downward, so the direction of weight-force is opposite to the direction of g-force acceleration:

$$\text{Weight} = \text{mass} \times -\text{g-force}$$

The reason for the minus sign is that the actual force (i.e., measured weight) on an object produced by a g-force is in the opposite direction to the sign of the g-force, since in physics, weight is not the force that produces the acceleration, but rather the equal-and-opposite reaction force to it. If the direction upward is taken as positive (the normal cartesian convention) then positive g-force (an acceleration vector that points upward) produces a force/weight on any mass, that acts downward (an example is positive-g acceleration of a rocket launch, producing downward weight). In the same way, a negative-g force is an acceleration vector downward (the negative direction on the y axis), and this acceleration downward produces a weight-force in a direction upward (thus pulling a pilot upward out of the seat, and forcing blood toward the head of a normally oriented pilot).

If a g-force (acceleration) is vertically upward and is applied by the ground (which is accelerating through space-time) or applied by the floor of an elevator to a standing person, most of the body experiences compressive stress which at any height, if multiplied by the area, is the related mechanical force, which is the product of the g-force and the supported mass (the mass above the level of support, including arms hanging down from above that level). At the same time, the arms themselves experience a tensile stress, which at any height, if multiplied by the area, is again the related mechanical force, which is the product of the g-force and the mass hanging below the point of mechanical support. The mechanical resistive force spreads from points of contact with the floor or supporting structure, and gradually decreases toward zero at the unsupported ends (the top in the case of support from below, such as a seat or the floor, the bottom for a hanging part of the body or object). With compressive force counted as negative tensile force, the rate of change of the tensile force in the direction of the g-force, per unit mass (the change between parts of the object such that the slice of the object between them has unit mass), is equal to the g-force plus the non-gravitational external forces on the slice, if any (counted positive in the direction opposite to the g-force).

For a given g-force the stresses are the same, regardless of whether this g-force is caused by mechanical resistance to gravity, or by a coordinate-acceleration (change in velocity) caused by a mechanical force, or by a combination of these. Hence, for people all mechanical forces feels exactly the same whether they cause coordinate acceleration or not. For objects likewise, the question of whether they can withstand the mechanical g-force without damage is the same for any type of g-force. For example, upward acceleration (e.g., increase of speed when going up or decrease of speed when going down) on Earth feels the same as being stationary on a celestial body with a higher surface gravity. Again, one should note that gravitation acting alone does not produce any g-force; g-force is only produced from mechanical pushes and pulls. For a free body (one that is free to move in space) such g-forces only arise as the "inertial" path that is the natural effect of gravitation, or the natural effect of the inertia of mass, is modified. Such modification may only arise from influences other than gravitation.

Examples of important situations involving g-forces include:

- The g-force acting on a stationary object resting on the Earth's surface is 1 g (upwards) and results from the resisting reaction of the Earth's surface bearing upwards equal to an acceleration of 1 g, and is equal and opposite to gravity. The number 1 is approximate, depending on location.
- The g-force acting on an object in any weightless environment such as free-fall in a vacuum is 0 g.

- The g-force acting on an object under acceleration can be much greater than 1 g, for example, the dragster pictured at top right can exert a horizontal g-force of 5.3 when accelerating.
- The g-force acting on an object under acceleration may be downwards, for example when cresting a sharp hill on a roller coaster.
- If there are no other external forces than gravity, the g-force in a rocket is the thrust per unit mass. Its magnitude is equal to the thrust-to-weight ratio times g, and to the consumption of delta-v per unit time.
- In the case of a shock, e.g., a collision, the g-force can be very large during a short time.

A classic example of negative g-force is in a fully inverted roller coaster which is accelerating (changing velocity) toward the ground. In this case, the roller coaster riders are accelerated toward the ground faster than gravity would accelerate them, and are thus pinned upside down in their seats. In this case, the mechanical force exerted by the seat causes the g-force by altering the path of the passenger downward in a way that differs from gravitational acceleration. The difference in downward motion, now faster than gravity would provide, is caused by the push of the seat, and it results in a g-force toward the ground.

All "coordinate accelerations" (or lack of them), are described by Newton's laws of motion as follows:

The Second Law of Motion, the law of acceleration states that: F = ma., meaning that a force F acting on a body is equal to the mass m of the body times its acceleration a.

The Third Law of Motion, the law of reciprocal actions states that: all forces occur in pairs, and these two forces are equal in magnitude and opposite in direction. Newton's third law of motion means that not only does gravity behave as a force acting downwards on, say, a rock held in your hand but also that the rock exerts a force on the Earth, equal in magnitude and opposite in direction.

This acrobatic airplane is pulling up in a +g maneuver; the pilot is experiencing several g's of inertial acceleration in addition to the force of gravity. The cumulative vertical axis forces acting upon his body make him momentarily 'weigh' many times more than normal.

In an airplane, the pilot's seat can be thought of as the hand holding the rock, the pilot as the rock. When flying straight and level at 1 g, the pilot is acted upon by the force of gravity. His weight (a downward force) is 725 newtons (163 lb_f). In accordance with Newton's third law, the plane and the seat underneath the pilot provides an equal and opposite force pushing upwards with a force of 725 N (163 lb_f). This mechanical force provides the 1.0 g-force upward proper acceleration on the pilot, even though this velocity in the upward direction does not change (this is similar to the situation of a person standing on the ground, where the ground provides this force and this g-force).

If the pilot were suddenly to pull back on the stick and make his plane accelerate upwards at 9.8 m/s^2, the total gforce on his body is 2 g, half of which comes from the seat pushing the pilot to resist gravity, and half from the seat pushing the pilot to cause his upward acceleration—a change in velocity which also is a proper acceleration because it also differs from a free fall trajectory. Considered in the frame of reference of the plane his body is now generating a force of 1,450 N (330 lb_f) downwards into his seat and the seat is simultaneously pushing upwards with an equal force of 1,450 N (330 lb_f).

Unopposed acceleration due to mechanical forces, and consequentially g-force, is experienced whenever anyone rides in a vehicle because it always causes a proper acceleration, and (in the absence of gravity) also always a coordinate acceleration (where velocity changes). Whenever the vehicle changes either direction or speed, the occupants feel lateral (side to side) or longitudinal (forward and backwards) forces produced by the mechanical push of their seats.

The expression "1 g = 9.80665 m/s^2" means that for every second that elapses, velocity changes 9.80665 meters per second (≡35.30394 km/h). This rate of change in velocity can also be denoted as 9.80665 (meter per second) per second, or 9.80665 m/s^2. For example: An acceleration of 1 g equates to a rate of change in velocity of approximately 35 kilometres per hour (22 mph) for each second that elapses. Therefore, if an automobile is capable of braking at 1 g and is traveling at 35 kilometres per hour (22 mph) it can brake to a standstill in one second and the driver will experience a deceleration of 1 g. The automobile traveling at three times this speed, 105 km/h (65 mph), can brake to a standstill in three seconds.

In the case of an increase in speed from 0 to v with constant acceleration within a distance of s this acceleration is $v^2/(2s)$.

Preparing an object for g-tolerance (not getting damaged when subjected to a high g-force) is called g-hardening. This may apply to, e.g., instruments in a projectile shot by a gun.

Human Tolerance

Human tolerances depend on the magnitude of the g-force, the length of time it is applied, the direction it acts, the location of application, and the posture of the body.

The human body is flexible and deformable, particularly the softer tissues. A hard slap on the face may briefly impose hundreds of g locally but not produce any real damage; a constant 16 g for a minute, however, may be deadly. When vibration is experienced, relatively low peak g levels can be severely damaging if they are at the resonance frequency of organs and connective tissues.

To some degree, g-tolerance can be trainable, and there is also considerable variation in innate ability between individuals. In addition, some illnesses, particularly cardiovascular problems, reduce g-tolerance.

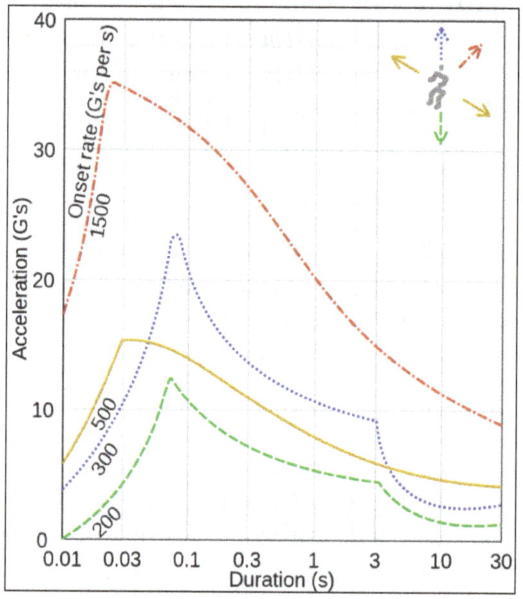

Semilog graph of the limits of tolerance of humans to linear acceleration.

Vertical

Aircraft pilots (in particular) sustain g-forces along the axis aligned with the spine. This causes significant variation in blood pressure along the length of the subject's body, which limits the maximum g-forces that can be tolerated.

Positive, or "upward" g, drives blood downward to the feet of a seated or standing person (more naturally, the feet and body may be seen as being driven by the upward force of the floor and seat, upward around the blood). Resistance to positive g varies. A typical person can handle about 5 g_0 (49 m/s^2) (meaning some people might pass out when riding a higher-g roller coaster, which in some cases exceeds this point) before losing consciousness, but through the combination of special g-suits and efforts to strain muscles—both of which act to force blood back into the brain—modern pilots can typically handle a sustained 9 g_0 (88 m/s^2).

In aircraft particularly, vertical g-forces are often positive (force blood towards the feet and away from the head); this causes problems with the eyes and brain in particular. As positive vertical g-force is progressively increased (such as in a centrifuge) the following symptoms may be experienced:

- Grey-out, where the vision loses hue, easily reversible on levelling out.
- Tunnel vision, where peripheral vision is progressively lost.
- Blackout, a loss of vision while consciousness is maintained, caused by a lack of blood to the head.
- G-LOC, a g-force induced loss of consciousness.

- Death, if g-forces are not quickly reduced, death can occur.

Resistance to "negative" or "downward" g, which drives blood to the head, is much lower. This limit is typically in the −2 to −3 g_0 (−20 to −29 m/s²) range. This condition is sometimes referred to as red out where vision is figuratively reddened due to the blood laden lower eyelid being pulled into the field of vision Negative g is generally unpleasant and can cause damage. Blood vessels in the eyes or brain may swell or burst under the increased blood pressure, resulting in degraded sight or even blindness.

Horizontal

The human body is better at surviving g-forces that are perpendicular to the spine. In general when the acceleration is forwards (subject essentially lying on their back, colloquially known as "eyeballs in") a much higher tolerance is shown than when the acceleration is backwards (lying on their front, "eyeballs out") since blood vessels in the retina appear more sensitive in the latter direction.

Early experiments showed that untrained humans were able to tolerate a range of accelerations depending on the time of exposure. This ranged from as much as 20 g for less than 10 seconds, to 10 g for 1 minute, and 6 g for 10 minutes for both eyeballs in and out. These forces were endured with cognitive facilities intact, as subjects were able to perform simple physical and communication tasks. The tests were determined to not cause long or short term harm although tolerance was quite subjective, with only the most motivated non-pilots capable of completing tests. The record for peak experimental horizontal g-force tolerance is held by acceleration pioneer John Stapp, in a series of rocket sled deceleration experiments culminating in a late 1954 test in which he was clocked in a little over a second from a land speed of Mach 0.9. He survived a peak "eyeballs-out" acceleration of 46.2 times the acceleration of gravity, and more than 25 g for 1.1 seconds, proving that the human body is capable of this. Stapp lived another 45 years to age 89 without any ill effects.

Short Duration Shock, Impact, and Jerk

Impact and mechanical shock are usually used to describe a high kinetic energy, short term excitation. A shock pulse is often measured by its peak acceleration in g-s and the pulse duration. Vibration is a periodic oscillation which can also be measured in g-s as well as frequency. The dynamics of these phenomena are what distinguish them from the g-forces caused by a relatively longer term accelerations.

After a free fall from a height the shock on an object during impact is g, where is the distance covered during the impact. For example, a stiff and compact object dropped from 1 m that impacts over a distance of 1 mm is subjected to a 1000 g deceleration.

Jerk is the rate of change of acceleration. In SI units, jerk is expressed as m/s³; it can also be expressed in standard gravity per second (g/s; 1 g/s ≈ 9.81 m/s³).

Other Biological Responses

Recent research carried out on extremophiles in Japan involved a variety of bacteria including E. coli and Paracoccus denitrificans being subject to conditions of extreme gravity. The bacteria were cultivated while being rotated in an ultracentrifuge at high speeds corresponding to 403,627 g. Paracoccus denitrificans was one of the bacteria which displayed not only survival but also robust cellular growth under these conditions of hyperacceleration which are usually only to be found in cosmic environments, such as on very massive stars or in the shock waves of supernovas. Analysis showed that the small size of prokaryotic cells is essential for successful growth under hypergravity. The research has implications on the feasibility of panspermia.

Typical Examples

Example	g-force*
The gyro rotors in Gravity Probe B and the free-floating proof masses in the TRIAD I navigation satellite	0 g
A ride in the Vomit Comet (parabolic flight)	≈ 0 g
Standing on the Moon at its equator	0.1654 g
Standing on the Earth at sea level–standard	1 g
Saturn V moon rocket just after launch	1.14 g
Bugatti Veyron from 0 to 100 km/h in 2.4 s	1.55 g†
Space Shuttle, maximum during launch and reentry	3 g
Gravitron amusement ride	2.5-3 g
High-g roller coasters	3.5–6.3 g
Top Fuel drag racing world record of 4.4 s over 1/4 mile	4.2 g
First World War Aircraft Sopwith Camel, Fokker D.VII, Fokker Dr.1, SPAD S.XIII, Nieuport 17 in a steep dive or back or front looping.	4.5–7 g
Formula One car, maximum under heavy braking	5.4 g
Formula One car, peak lateral in turns	5–6 g
Luge, maximum expected at the Whistler Sliding Centre	5.2 g
Standard, full aerobatics certified glider	+7/–5 g
Apollo 16 on reentry	7.19 g
Maximum permitted g-force turn in Red Bull Air Race plane	10g
Maximum for human on a rocket sled	46.2 g
Sprint missile	100 g
Brief human exposure survived in crash	> 100 g
Space gun with a barrel length of 1 km and a muzzle velocity of 6 km/s, as proposed by Quicklaunch (assuming constant acceleration)	1,800 g
Shock capability of mechanical wrist watches	> 5,000 g
V8 Formula One engine, maximum piston acceleration	8,600 g
Rating of electronics built into military artillery shells	15,500 g
Analytical ultracentrifuge spinning at 60,000 rpm, at the bottom of the analysis cell (7.2 cm)	300,000 g
Mean acceleration of a proton in the Large Hadron Collider	190,000,000 g
Gravitational acceleration at the surface of a typical neutron star	2.0×10^{11} g
Acceleration from a wakefield plasma accelerator	8.9×10^{20} g

* Including contribution from resistance to gravity.

† Directed 40 degrees from horizontal.

Measurement using an Accelerometer

The Superman: Escape from Krypton roller coaster at Six Flags Magic Mountain provides 6.5 seconds of ballistic weightlessness.

An accelerometer, in its simplest form, is a damped mass on the end of a spring, with some way of measuring how far the mass has moved on the spring in a particular direction, called an 'axis'.

Accelerometers are often calibrated to measure g-force along one or more axes. If a stationary, single-axis accelerometer is oriented so that its measuring axis is horizontal, its output will be 0 g, and it will continue to be 0 g if mounted in an automobile traveling at a constant velocity on a level road. When the driver presses on the brake or gas pedal, the accelerometer will register positive or negative acceleration.

If the accelerometer is rotated by 90° so that it is vertical, it will read +1 g upwards even though stationary. In that situation, the accelerometer is subject to two forces: the gravitational force and the ground reaction force of the surface it is resting on. Only the latter force can be measured by the accelerometer, due to mechanical interaction between the accelerometer and the ground. The reading is the acceleration the instrument would have if it were exclusively subject to that force.

A three-axis accelerometer will output zerog on all three axes if it is dropped or otherwise put into a ballistic trajectory (also known as an inertial trajectory), so that it experiences "free fall," as do astronauts in orbit (astronauts experience small tidal accelerations called microgravity, which are neglected for the sake of discussion here). Some amusement park rides can provide several seconds at near-zero g. Riding NASA's "Vomit Comet" provides near-zero g for about 25 seconds at a time.

Fuel Efficiency

Fuel efficiency is a form of thermal efficiency, meaning the efficiency of a process that converts chemical potential energy contained in a carrier fuel into kinetic energy or work. Overall fuel efficiency may vary per device, which in turn may vary per application fuel efficiency, especially fossil fuel power plants or industries dealing with combustion, such as ammonia production during the Haber process.

In the context of transport, fuel economy is the energy efficiency of a particular vehicle, given as a ratio of distance traveled per unit of fuel consumed. It is dependent on engine efficiency, transmission design, and tire design. Fuel economy is expressed in miles per gallon (mpg) in the USA and usually also in the UK (imperial gallon); there is sometimes confusion as the imperial gallon is 20% larger than the US gallon so that mpg values are not directly comparable. In countries using the metric system fuel economy is stated as "fuel consumption" in liters per 100 kilometers (L/100 km). Litres per mil are used in Norway and Sweden.

Fuel consumption is a more accurate measure of a vehicle's performance because it is a linear relationship while fuel economy leads to distortions in efficiency improvements.

Weight-specific efficiency (efficiency per unit weight) may be stated for freight, and passenger-specific efficiency (vehicle efficiency per passenger).

Vehicle Design

Fuel efficiency is dependent on many parameters of a vehicle, including its engine parameters, aerodynamic drag, weight, and rolling resistance. There have been advances in all areas of vehicle design in recent decades.

Hybrid vehicles use two or more power sources for propulsion. In many designs, a small combustion engine is combined with electric motors. Kinetic energy which would otherwise be lost to heat during braking is recaptured as electrical power to improve fuel efficiency. Engines automatically shut off when vehicles come to a stop and start again when the accelerator is pressed preventing wasted energy from idling.

Fleet Efficiency

Fleet efficiency describes the average efficiency of a population of vehicles. Technological advances in efficiency may be offset by a change in buying habits with a propensity to heavier vehicles, which are less efficient, all else being equal.

Energy Efficiency Terminology

Energy efficiency is similar to fuel efficiency but the input is usually in units of energy such as British thermal units (BTU), megajoules (MJ), gigajoules (GJ), kilocalories (kcal), or kilowatt-hours (kW·h). The inverse of "energy efficiency" is "energy intensity", or the amount of input energy required for a unit of output such as MJ/passenger-km (of passenger transport), BTU/ton-mile or kJ/t-km (of freight transport), GJ/t (for production of steel and other materials), BTU/(kW·h) (for electricity generation), or litres/100 km (of vehicle travel). Litres per 100 km is also a measure of "energy intensity" where the input is measured by the amount of fuel and the output is measured by the distance travelled. For example: Fuel economy in automobiles.

Given a heat value of a fuel, it would be trivial to convert from fuel units (such as litres of gasoline) to energy units (such as MJ) and conversely. But there are two problems with comparisons made using energy units:

- There are two different heat values for any hydrogen-containing fuel which can differ by several percent.
- When comparing transportation energy costs, it must be remembered that a kilowatt hour of electric energy may require an amount of fuel with heating value of 2 or 3 kilowatt hours to produce it.

Energy Content of Fuel

The specific energy content of a fuel is the heat energy obtained when a certain quantity is burned (such as a gallon, litre, kilogram). It is sometimes called the heat of combustion. There exists two different values of specific heat energy for the same batch of fuel. One is the high (or gross) heat of combustion and the other is the low (or net) heat of combustion. The high value is obtained when, after the combustion, the water in the exhaust is in liquid form. For the low value, the exhaust has all the water in vapor form (steam). Since water vapor gives up heat energy when it changes from vapor to liquid, the liquid water value is larger since it includes the latent heat of vaporization of water. The difference between the high and low values is significant, about 8 or 9%. This accounts for most of the apparent discrepancy in the heat value of gasoline. In the U.S. (and the table) the high heat values have traditionally been used, but in many other countries, the low heat values are commonly used.

Fuel type	MJ/L	MJ/kg	BTU/imp gal	BTU/US gal	Research octane number (RON)
Regular gasoline/petrol	34.8	~47	150,100	125,000	Min. 91
Premium gasoline/petrol		~46			Min. 95
Autogas (LPG) (60% propane and 40% butane)	25.5–28.7	~51			108–110
Ethanol	23.5	31.1	101,600	84,600	129
Methanol	17.9	19.9	77,600	64,600	123
Gasohol (10% ethanol and 90% gasoline)	33.7	~45	145,200	121,000	93/94
E85 (85% ethanol and 15% gasoline)	25.2	~33	108,878	90,660	100–105
Diesel	38.6	~48	166,600	138,700	N/A
Biodiesel	35.1	39.9	151,600	126,200	N/A
Vegetable oil (using 9.00 kcal/g)	34.3	37.7	147,894	123,143	
Aviation gasoline	33.5	46.8	144,400	120,200	80-145
Jet fuel, naphtha	35.5	46.6	153,100	127,500	N/A to turbine engines
Jet fuel, kerosene	37.6	~47	162,100	135,000	N/A to turbine engines
Liquefied natural gas	25.3	~55	109,000	90,800	
Liquid hydrogen	9.3	~130	40,467	33,696	

Neither the gross heat of combustion nor the net heat of combustion gives the theoretical amount of mechanical energy (work) that can be obtained from the reaction. (This is given by the change in Gibbs free energy, and is around 45.7 MJ/kg for gasoline.) The actual amount of mechanical work obtained from fuel (the inverse of the specific fuel consumption) depends on the engine. A figure of 17.6 MJ/kg is possible with a gasoline engine, and 19.1 MJ/kg for a diesel engine.

Fuel Efficiency of Motor Vehicles

The fuel efficiency of motor vehicles can be expressed in more ways:

- Fuel consumption is the amount of fuel used per unit distance; for example, litres per 100 kilometres (L/100 km). In this case, the lower the value, the more economic a vehicle is (the less fuel it needs to travel a certain distance); this is the measure generally used across Europe (except the UK, Denmark and The Netherlands), New Zealand, Australia and Canada. Also in Uruguay, Paraguay, Guatemala, Colombia, China, and Madagascar., as also in post-Soviet space.

- Fuel economy is the distance travelled per unit volume of fuel used; for example, kilometres per litre (km/L) or miles per gallon (MPG), where 1 MPG (imperial) ≈ 0.354006 km/L. In this case, the higher the value, the more economic a vehicle is (the more distance it can travel with a certain volume of fuel). This measure is popular in the USA and the UK (mpg), but in Europe, India, Japan, South Korea and Latin America the metric unit km/L is used instead.

Converting from mpg or to L/100 km (or vice versa) involves the use of the reciprocal function, which is not distributive. Therefore, the average of two fuel economy numbers gives different values if those units are used, because one of the functions is reciprocal, thus not linear. If two people calculate the fuel economy average of two groups of cars with different units, the group with better fuel economy may be one or the other. However, from the point of energy used as a shared method of measure, the result shall be the same in both the cases.

In parts of Europe, the two standard measuring cycles for "litre/100 km" value are "urban" traffic with speeds up to 50 km/h from a cold start, and then "extra urban" travel at various speeds up to 120 km/h which follows the urban test. A combined figure is also quoted showing the total fuel consumed in divided by the total distance traveled in both tests. A reasonably modern European supermini and many mid-size cars, including station wagons, may manage motorway travel at 5 L/100 km (47 mpg US/56 mpg imp) or 6.5 L/100 km in city traffic (36 mpg US/43 mpg imp), with carbon dioxide emissions of around 140 g/km.

An average North American mid-size car travels 21 mpg (US) (11 L/100 km) city, 27 mpg (US) (9 L/100 km) highway; a full-size SUV usually travels 13 mpg (US) (18 L/100 km) city and 16 mpg (US) (15 L/100 km) highway. Pickup trucks vary considerably; whereas a 4 cylinder-engined light pickup can achieve 28 mpg (8 L/100 km), a V8 full-size pickup with extended cabin only travels 13 mpg (US) (18 L/100 km) city and 15 mpg (US) (15 L/100 km) highway.

The average fuel economy is higher in Europe due to the higher cost of fuel. In the UK, a gallon of gas without tax would cost US$1.97, but with taxes cost US$6.06 in 2005. The average cost in the United States was US$2.61. Consumers prefer "muscle cars" but choose more fuel efficient ones when gas prices increase.

European-built cars are generally more fuel-efficient than US vehicles. While Europe has many higher efficiency diesel cars, European gasoline vehicles are on average also more efficient than gasoline-powered vehicles in the USA. Most European vehicles cited in the CSI study run on diesel engines, which tend to achieve greater fuel efficiency than gas engines. Selling those cars in the United States is difficult because of emission standards, notes Walter McManus, a fuel economy expert at the University of Michigan Transportation Research Institute. "For the most part, European diesels don't meet U.S. emission standards", McManus said in 2007. Another reason why many European models are not marketed in the United States is that labor unions object to having the big 3 import any new foreign built models regardless of fuel economy while laying off workers at home.

An example of European cars' capabilities of fuel economy is the microcar Smart Fortwo cdi, which can achieve up to 3.4 l/100 km (69.2 mpg US) using a turbocharged three-cylinder 41 bhp (30 kW) Diesel engine. The Fortwo is produced by Daimler AG and is currently only sold by one company in the United States. Furthermore, the current (and to date already 10-year-old) world record in fuel economy of production cars is held by the Volkswagen Group, with special production models (labeled "3L") of the Volkswagen Lupo and the Audi A2, consuming as little as 3 L/100 km (94 mpg_{-imp}; 78 mpg_{-US}).

Diesel engines generally achieve greater fuel efficiency than petrol (gasoline) engines. Passenger car diesel engines have energy efficiency of up to 41% but more typically 30%, and petrol engines of up to 37.3%, but more typically 20%. That is one of the reasons why diesels have better fuel efficiency than equivalent petrol cars. A common margin is 25% more miles per gallon for an efficient turbodiesel.

For example, the current model Skoda Octavia, using Volkswagen engines, has a combined European fuel efficiency of 41.3 mpg for the 105 bhp (78 kW) petrol engine and 52.3 mpg for the 105 bhp (78 kW) — and heavier — diesel engine. The higher compression ratio is helpful in raising the energy efficiency, but diesel fuel also contains approximately 10% more energy per unit volume than gasoline which contributes to the reduced fuel consumption for a given power output.

In 2002, the United States had 85,174,776 trucks, and averaged 13.5 miles per US gallon (17.4 L/100 km; 16.2 mpg_{-imp}). Large trucks, over 33,000 pounds (15,000 kg), averaged 5.7 miles per US gallon (41 L/100 km; 6.8 mpg_{-imp}).

Truck fuel economy					
GVWR lbs	Number	Percentage	Average miles per truck	fuel economy	Percentage of fuel use
6,000 lbs and less	51,941,389	61.00%	11,882	17.6	42.70%
6,001 – 10,000 lbs	28,041,234	32.90%	12,684	14.3	30.50%
Light truck subtotal	79,982,623	93.90%	12,163	16.2	73.20%
10,001 – 14,000 lbs	691,342	0.80%	14,094	10.5	1.10%
14,001 – 16,000 lbs	290,980	0.30%	15,441	8.5	0.50%

16,001 – 19,500 lbs	166,472	0.20%	11,645	7.9	0.30%
19,501 – 26,000 lbs	1,709,574	2.00%	12,671	7	3.20%
Medium truck subtotal	2,858,368	3.40%	13,237	8	5.20%
26,001 – 33,000 lbs	179,790	0.20%	30,708	6.4	0.90%
33,001 lbs and up	2,153,996	2.50%	45,739	5.7	20.70%
Heavy truck subtotal	2,333,786	2.70%	44,581	5.8	21.60%
Total	85,174,776	100.00%	13,088	13.5	100.00%

The average economy of automobiles in the United States in 2002 was 22.0 miles per US gallon (10.7 L/100 km; 26.4 mpg_{-imp}). By 2010 this had increased to 23.0 miles per US gallon (10.2 L/100 km; 27.6 mpg_{-imp}). Average fuel economy in the United States gradually declined until 1973, when it reached a low of 13.4 miles per US gallon (17.6 L/100 km; 16.1 mpg_{-imp}) and gradually has increased since, as a result of higher fuel cost. A study indicates that a 10% increase in gas prices will eventually produce a 2.04% increase in fuel economy. One method by car makers to increase fuel efficiency is lightweighting in which lighter-weight materials are substituted in for improved engine performance and handling.

Fuel Efficiency in Microgravity

How fuel combusts affects how much energy is produced. The National Aeronautics and Space Administration (NASA) has investigated fuel consumption in microgravity.

The common distribution of a flame under normal gravity conditions depends on convection, because soot tends to rise to the top of a flame, such as in a candle, making the flame yellow. In microgravity or zero gravity, such as an environment in outer space, convection no longer occurs, and the flame becomes spherical, with a tendency to become more blue and more efficient. There are several possible explanations for this difference, of which the most likely one given is the hypothesis that the temperature is evenly distributed enough that soot is not formed and complete combustion occurs., National Aeronautics and Space Administration, April 2005. Experiments by NASA in microgravity reveal that diffusion flames in microgravity allow more soot to be completely oxidised after they are produced than diffusion flames on Earth, because of a series of mechanisms that behaved differently in microgravity when compared to normal gravity conditions.LSP-1 experiment results, National Aeronautics and Space Administration, April 2005. Premixed flames in microgravity burn at a much slower rate and more efficiently than even a candle on Earth, and last much longer.

Transportation

Vehicle Efficiency and Transportation Pollution

Fuel efficiency directly affects emissions causing pollution by affecting the amount of fuel used. However, it also depends on the fuel source used to drive the vehicle concerned. Cars for example, can run on a number of fuel types other than gasoline, such as natural gas, LPG or biofuel or electricity which creates various quantities of atmospheric pollution.

A kilogram of carbon, whether contained in petrol, diesel, kerosene, or any other hydrocarbon fuel in a vehicle, leads to approximately 3.6 kg of CO_2 emissions. Due to the carbon content of gasoline, its combustion emits 2.3 kg/l (19.4 lb/US gal) of CO_2; since diesel fuel is more energy dense per unit volume, diesel emits 2.6 kg/l (22.2 lb/US gal). This figure is only the CO_2 emissions of the final fuel product and does not include additional CO_2 emissions created during the drilling, pumping, transportation and refining steps required to produce the fuel. Additional measures to reduce overall emission includes improvements to the efficiency of air conditioners, lights and tires.

Driving Technique

Many drivers have the potential to improve their fuel efficiency significantly. These five basic fuel-efficient driving techniques can be effective. Simple things such as keeping tires properly inflated, having a vehicle well-maintained and avoiding idling can dramatically improve fuel efficiency.

There is a growing community of enthusiasts known as hypermilers who develop and practice driving techniques to increase fuel efficiency and reduce consumption. Hypermilers have broken records of fuel efficiency, for example, achieving 109 miles per gallon in a Prius. In non-hybrid vehicles these techniques are also beneficial, with fuel efficiencies of up to 59 MPG in a Honda Accord or 30 MPG in an Acura MDX.

Advanced Technology Improvements to Improve Fuel Efficiency

The most efficient machines for converting energy to rotary motion are electric motors, as used in electric vehicles. However, electricity is not a primary energy source so the efficiency of the electricity production has also to be taken into account. Currently railway trains can be powered using electricity, delivered through an additional running rail, overhead catenary system or by on-board generators used in diesel-electric locomotives as common on the US and UK rail networks. Pollution produced from centralised generation of electricity is emitted at a distant power station, rather than "on site". Pollution can be reduced by using more railway electrification and low carbon power for electricity. Some railways, such as the French SNCF and Swiss federal railways derive most, if not 100% of their power, from hydroelectric or nuclear power stations, therefore atmospheric pollution from their rail networks is very low. This was reflected in a study by AEA Technology between a Eurostar train and airline journeys between London and Paris, which showed the trains on average emitting 10 times less CO_2, per passenger, than planes, helped in part by French nuclear generation.

In the future, hydrogen cars may be commercially available. Toyota is test marketing hydrogen fuel cell powered vehicles in southern California where a series of hydrogen fueling stations has been established. Powered either through chemical reactions in a fuel cell that create electricity to drive very efficient electrical motors or by directly burning hydrogen in a combustion engine (near identically to a natural gas vehicle, and similarly compatible with both natural gas and gasoline); these vehicles promise to have near-zero pollution from the tailpipe (exhaust pipe). Potentially the atmospheric pollution could be minimal, provided the hydrogen is made by electrolysis using electricity from non-polluting sources such as solar, wind or hydroelectricity or nuclear. Commercial hydrogen production uses fossil fuels and produces more carbon dioxide than hydrogen.

Because there are pollutants involved in the manufacture and destruction of a car and the production, transmission and storage of electricity and hydrogen, the use of the label "zero pollution" should be understood as applying only to the car's conversion of stored energy into transportation.

In 2004, a consortium of major auto-makers — BMW, General Motors, Honda, Toyota and Volkswagen/Audi — came up with "Top Tier Detergent Gasoline Standard" to gasoline brands in the US and Canada that meet their minimum standards for detergent content and do not contain metallic additives. Top Tier gasoline contains higher levels of detergent additives in order to prevent the build-up of deposits (typically, on fuel injector and intake valve) known to reduce fuel economy and engine performance.

Safety Engineering

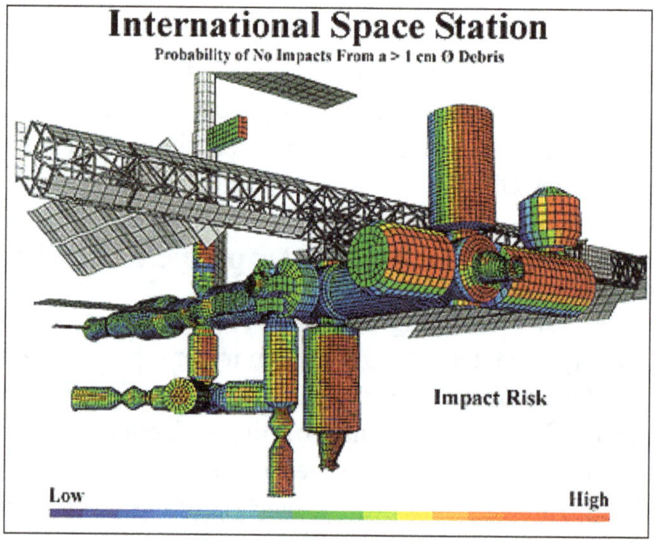

NASA's illustration showing high impact risk areas for the International Space Station.

Safety engineering is an engineering discipline which assures that engineered systems provide acceptable levels of safety. It is strongly related to industrial engineering/systems engineering, and the subset system safety engineering. Safety engineering assures that a life-critical system behaves as needed, even when components fail.

Analysis Techniques

Analysis techniques can be split into two categories: qualitative and quantitative methods. Both approaches share the goal of finding causal dependencies between a hazard on system level and failures of individual components. Qualitative approaches focus on the question "What must go wrong, such that a system hazard may occur?", while quantitative methods aim at providing estimations about probabilities, rates and/or severity of consequences.

The complexity of the technical systems such as Improvements of Design and Materials, Planned Inspections, Fool-proof design, and Backup Redundancy decreases risk and increases the cost. The risk can be decreased to ALARA (as low as reasonably achievable) or ALAPA (as low as practically achievable) levels.

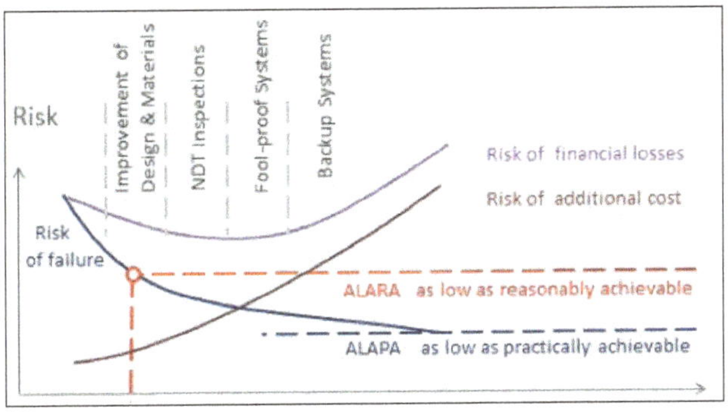

Risk vs Cost/Complexity.

Traditionally, safety analysis techniques rely solely on skill and expertise of the safety engineer. In the last decade model-based approaches have become prominent. In contrast to traditional methods, model-based techniques try to derive relationships between causes and consequences from some sort of model of the system.

Traditional Methods for Safety Analysis

The two most common fault modeling techniques are called failure mode and effects analysis and fault tree analysis. These techniques are just ways of finding problems and of making plans to cope with failures, as in probabilistic risk assessment. One of the earliest complete studies using this technique on a commercial nuclear plant was the WASH-1400 study, also known as the Reactor Safety Study or the Rasmussen Report.

Failure Modes and Effects Analysis

Failure Mode and Effects Analysis (FMEA) is a bottom-up, inductive analytical method which may be performed at either the functional or piece-part level. For functional FMEA, failure modes are identified for each function in a system or equipment item, usually with the help of a functional block diagram. For piece-part FMEA, failure modes are identified for each piece-part component (such as a valve, connector, resistor, or diode). The effects of the failure mode are described, and assigned a probability based on the failure rate and failure mode ratio of the function or component. This quantiazation is difficult for software ---a bug exists or not, and the failure models used for hardware components do not apply. Temperature and age and manufacturing variability affect a resistor; they do not affect software.

Failure modes with identical effects can be combined and summarized in a Failure Mode Effects Summary. When combined with criticality analysis, FMEA is known as Failure Mode, Effects, and Criticality Analysis or FMECA, pronounced "fuh-MEE-kuh".

Fault Tree Analysis

Fault tree analysis (FTA) is a top-down, deductive analytical method. In FTA, initiating primary events such as component failures, human errors, and external events are traced through Boolean logic gates to an undesired top event such as an aircraft crash or nuclear reactor core melt. The intent is to identify ways to make top events less probable, and verify that safety goals have been achieved.

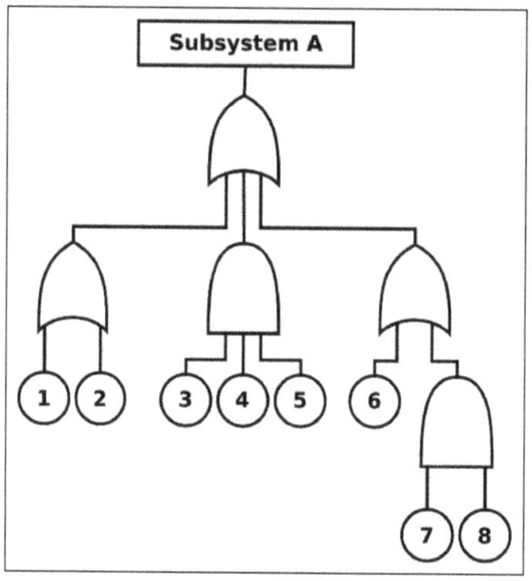

A fault tree diagram.

Fault trees are a logical inverse of success trees, and may be obtained by applying de Morgan's theorem to success trees (which are directly related to reliability block diagrams).

FTA may be qualitative or quantitative. When failure and event probabilities are unknown, qualitative fault trees may be analyzed for minimal cut sets. For example, if any minimal cut set contains a single base event, then the top event may be caused by a single failure. Quantitative FTA is used to compute top event probability, and usually requires computer software such as CAFTA from the Electric Power Research Institute or SAPHIRE from the Idaho National Laboratory.

Some industries use both fault trees and event trees. An event tree starts from an undesired initiator (loss of critical supply, component failure etc.) and follows possible further system events through to a series of final consequences. As each new event is considered, a new node on the tree is added with a split of probabilities of taking either branch. The probabilities of a range of "top events" arising from the initial event can then be seen.

Safety Certification

Usually a failure in safety-certified systems is acceptable if, on average, less than one life per 10^9 hours of continuous operation is lost to failure.{as per FAA document AC 25.1309-1A} Most Western nuclear reactors, medical equipment, and commercial aircraft are certified to this level. The cost versus loss of lives has been considered appropriate at this level (by FAA for aircraft systems under Federal Aviation Regulations).

Preventing Failure

Once a failure mode is identified, it can usually be mitigated by adding extra or redundant equipment to the system. For example, nuclear reactors contain dangerous radiation, and nuclear reactions can cause so much heat that no substance might contain them. Therefore, reactors have emergency core cooling systems to keep the temperature down, shielding to contain the radiation, and engineered barriers (usually several, nested, surmounted by a containment building) to prevent accidental leakage. Safety-critical systems are commonly required to permit no single event or component failure to result in a catastrophic failure mode.

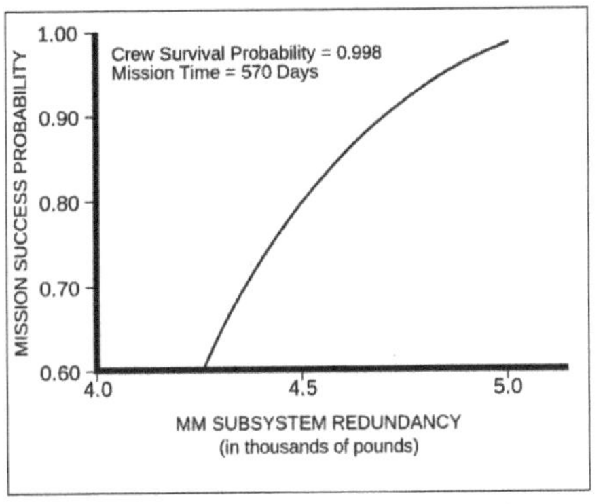

A NASA graph shows the relationship between the survival of a crew of astronauts and the amount of redundant equipment in their spacecraft (the "MM", Mission Module).

Most biological organisms have a certain amount of redundancy: multiple organs, multiple limbs, etc.

For any given failure, a fail-over or redundancy can almost always be designed and incorporated into a system.

There are two categories of techniques to reduce the probability of failure: Fault avoidance techniques increase the reliability of individual items (increased design margin, de-rating, etc.). Fault tolerance techniques increase the reliability of the system as a whole (redundancies, barriers, etc.).

Safety and Reliability

Safety engineering and reliability engineering have much in common, but safety is not reliability. If a medical device fails, it should fail safely; other alternatives will be available to the surgeon. If the engine on a single-engine aircraft fails, there is no backup. Electrical power grids are designed for both safety and reliability; telephone systems are designed for reliability, which becomes a safety issue when emergency (e.g. US "911") calls are placed.

Probabilistic risk assessment has created a close relationship between safety and reliability. Component reliability, generally defined in terms of component failure rate, and external event probability are both used in quantitative safety assessment methods such as FTA. Related probabilistic methods are used to determine system Mean Time Between Failure (MTBF), system availability, or probability of mission success or failure. Reliability analysis has a broader scope than safety analysis, in that non-critical failures are considered. On the other hand, higher failure rates are considered acceptable for non-critical systems.

Safety generally cannot be achieved through component reliability alone. Catastrophic failure probabilities of 10^{-9} per hour correspond to the failure rates of very simple components such as resistors or capacitors. A complex system containing hundreds or thousands of components might be able to achieve a MTBF of 10,000 to 100,000 hours, meaning it would fail at 10^{-4} or 10^{-5} per hour. If a system failure is catastrophic, usually the only practical way to achieve 10^{-9} per hour failure rate is through redundancy.

When adding equipment is impractical (usually because of expense), then the least expensive form of design is often "inherently fail-safe". That is, change the system design so its failure modes are not catastrophic. Inherent fail-safes are common in medical equipment, traffic and railway signals, communications equipment, and safety equipment.

The typical approach is to arrange the system so that ordinary single failures cause the mechanism to shut down in a safe way (for nuclear power plants, this is termed a passively safe design, although more than ordinary failures are covered). Alternately, if the system contains a hazard source such as a battery or rotor, then it may be possible to remove the hazard from the system so that its failure modes cannot be catastrophic. The U.S. Department of Defense Standard Practice for System Safety (MIL–STD–882) places the highest priority on elimination of hazards through design selection.

One of the most common fail-safe systems is the overflow tube in baths and kitchen sinks. If the valve sticks open, rather than causing an overflow and damage, the tank spills into an overflow. Another common example is that in an elevator the cable supporting the car keeps spring-loaded brakes open. If the cable breaks, the brakes grab rails, and the elevator cabin does not fall.

Some systems can never be made fail safe, as continuous availability is needed. For example, loss of engine thrust in flight is dangerous. Redundancy, fault tolerance, or recovery procedures are used for these situations (e.g. multiple independent controlled and fuel fed engines). This also makes the system less sensitive for the reliability prediction errors or quality induced uncertainty for the separate items. On the other hand, failure detection & correction and avoidance of common cause failures becomes here increasingly important to ensure system level reliability.

Engine Power

Engine power or horsepower is the maximum power that an engine can put out. It can be expressed in kilowatts or horsepower. The power output depends on the size and design of the engine, but also on the speed at which it is running and the load or torque. Maximum power is achieved at relatively high speeds and at high load.

Examples

Various Engines

The following tables give some examples of engine power for a wide variety of engines and vehicles.

Heat Engine/Heat pump type	Peak Power Output		Power-to-weight ratio		Example Use
Wärtsilä RTA96-C 14-cylinder two-stroke Turbo Diesel engine	80,080 kW	108,920 hp	0.03 kW/kg	0.02 hp/lb	Emma Mærsk container ship
Suzuki 538 cc V2 4-stroke gas (petrol) outboard Otto engine	19 kW	25 hp	0.27 kW/kg	0.16 hp/lb	Runabout boats
DOE/NASA/0032-28 Mod 2 502 cc gas (petrol) Stirling engine	62.3 kW	83.5 hp	0.30 kW/kg	0.18 hp/lb	Chevrolet Celebrity[•] 1985
GM 6.6 L Duramax LMM (LYE option) V8 Turbo Diesel engine	246 kW	330 hp	0.65 kW/kg	0.40 hp/lb	Chevrolet Kodiak[•], GMC Topkick[•]

Junkers Jumo 205A opposed-piston two-stroke Diesel engine	647 kW	867 hp	1.1 kW/kg	0.66 hp/lb	Ju 86C-1 airliner, B&V Ha 139 floatplane
GE LM2500+ marine turboshaft Brayton gas turbine	30,200 kW	40,500 hp	1.31 kW/kg	0.80 hp/lb	GTS Millennium cruiseship, QM2 ocean liner
Mazda 13B-MSP Renesis 1.3 L Wankel engine	184 kW	247 hp	1.5 kW/kg	0.92 hp/lb	Mazda RX-8[•]
PW R-4360 71.5 L 28-cylinder supercharged Radial engine	3,210 kW	4,300 hp	1.83 kW/kg	1.11 hp/lb	B-50 Superfortress, Convair B-36
					C-97 Stratofreighter, C-119 Flying Boxcar
					Hughes H-4 Hercules "Spruce Goose"
Wright R-3350 54.57 L 18-c s/c Turbo-compound Radial engine	2,535 kW	3,400 hp	2.09 kW/kg	1.27 hp/lb	B-29 Superfortress, Douglas DC-7
					C-97 S/f prototype, Kaiser-Frazer C-119F
O.S. Engines 49-PI Type II 4.97 cc UAV Wankel engine	0.934 kW	1.252 hp	2.8 kW/kg	1.7 hp/lb	Model aircraft, Radio-controlled aircraft
GE LM6000 marine turboshaft Brayton gas turbine	44,700 kW	59,900 hp	5.67 kW/kg	3.38 hp/lb	Peaking power plant
GE CF6-80C2 Brayton high-bypass turbofan jet engine					Boeing 747[•], 767, Airbus A300
BMW V10 3L P84/5 2005 gas (petrol) Otto engine	690 kW	925 hp	7.5 kW/kg	4.6 hp/lb	Williams FW27 car[•], Formula One auto racing
GE90-115B Brayton turbofan jet engine	83,164 kW	111,526 hp	10.0 kW/kg	6.10 hp/lb	Boeing 777
PWR RS-24 (SSME) Block II H_2 Brayton turbopump	63,384 kW	85,000 hp	138 kW/kg	84 hp/lb	Space Shuttle (STS-110 and later) [•]
PWR RS-24 (SSME) Block I H_2 Brayton turbopump	53,690 kW	72,000 hp	153 kW/kg	93 hp/lb	Space Shuttle

Notable Low Ratio, (Listed as Weight to Power)

Vehicle	Power	Vehicle Weight	Weight to Power ratio
Benz Patent Motorwagen 954 cc 1886	560 W / 0.75 bhp	265 kg / 584 lb	2.1 W/kg / 779 lb/hp
Stephenson's Rocket 0-2-2 steam locomotive with tender 1829	15 kW / 20 bhp	4,320 kg / 9524 lb	3.5 W/kg / 476 lb/hp
CBQ Zephyr streamliner diesel locomotive with railcars 1934	492 kW / 660 bhp	94 t / 208,000 lb	5.21 W/kg / 315 lb/hp
Alberto Contador's Verbier climb 2009 Tour de France on Specialized bike	420 W / 0.56 bhp	62 kg / 137 lb	6.7 W/kg / 245 lb/hp
Force Motors Minidor Diesel 499 cc auto rickshaw	6.6 kW / 8.8 bhp	700 kg / 1543 lb	9 W/kg / 175 lb/hp

PRR Q2 4-4-6-4 steam locomotive with tender 1944	5,956 kW / 7,987 bhp	475.9 t / 1,049,100 lb	12.5 W/kg / 131 lb/hp
Mercedes-Benz Citaro O530BZ H_2 fuel cell bus 2002	205 kW / 275 bhp	14,500 kg / 32,000 lb	14.1 W/kg / 116 lb/hp
TGV BR Class 373 high-speed Eurostar Trainset 1993	12,240 kW / 16,414 bhp	816 t / 1,798,972 lb	15 W/kg / 110 lb/hp
General Dynamics M1 Abrams Main battle tank 1980	1,119 kW / 1500 bhp	55.7 t / 122,800 lb	20.1 W/kg / 81.9 lb/hp
BR Class 43 high-speed diesel electric locomotive 1975	1,678 kW / 2,250 bhp	70.25 t / 154,875 lb	23.9 W/kg / 69 lb/hp
GE AC6000CW diesel electric locomotive 1996	4,660 kW / 6,250 bhp	192 t / 423,000 lb	24.3 W/kg / 68 lb/hp
BR Class 55 Napier Deltic diesel electric locomotive 1961	2,460 kW / 3,300 bhp	101 t / 222,667 lb	24.4 W/kg / 68 lb/hp
International CXT 2004	164 kW / 220 bhp	6,577 kg / 14500 lb	25 W/kg / 66 lb/hp
Ford Model T 2.9 L flex-fuel 1908	15 kW / 20 bhp	540 kg / 1,200 lb	28 W/kg / 60 lb/hp
TH!NK City 2008	30 kW / 40 bhp	1038 kg / 2,288 lb	28.9 W/kg / 56.9 lb/hp
Messerschmitt KR200 Kabinenroller 191 cc 1955	6 kW / 8.2 bhp	230 kg / 506 lb	30 W/kg / 50 lb/hp
Wright Flyer 1903	9 kW / 12 bhp	274 kg / 605 lb	33 W/kg / 50 lb/hp
Tata Nano 624 cc 2008	26 kW / 35 bhp	635 kg / 1,400 lb	41.0 W/kg / 40 lb/hp
Bombardier JetTrain high-speed gas turbine-electric locomotive 2000	3,750 kW / 5,029 bhp	90,750 kg / 200,000 lb	41.2 W/kg / 39.8 lb/hp
Suzuki MightyBoy 543 cc 1988	23 kW / 31 bhp	550 kg / 1,213 lb	42 W/kg / 39 lb/hp
Mitsubishi i MiEV 2009	47 kW / 63 bhp	1,080 kg / 2,381 lb	43.5 W/kg / 37.8 lb/hp
Holden FJ 2,160 cc 1953	44.7 kW / 60 bhp	1,021 kg / 2,250 lb	43.8 W/kg / 37.5 lb/hp
Chevrolet Kodiak/GMC Topkick LYE 6.6 L 2005	246 kW / 330 bhp	5126 kg / 11,300 lb	48 W/kg / 34.2 lb/hp
DOE/NASA/0032-28 Chevrolet Celebrity 502 cc ASE Mod II 1985	62.3 kW / 83.5 bhp	1,297 kg / 2,860 lb	48.0 W/kg / 34.3 lb/hp
Suzuki Alto 796 cc 2000	35 kW / 46 bhp	720 kg / 1,587 lb	49 W/kg / 35 lb/hp
Land Rover Defender 2.4 L 1990	90 kW / 121 bhp	1,837 kg / 4,050 lb	49 W/kg / 33 lb/hp

Common Power, (Listed as Weight to Power)

Vehicle	Power	Vehicle Weight	Weight-to-Power ratio
Toyota Prius 1.8 L 2010 (petrol only)	73 kW / 98 bhp	1,380 kg / 3,042 lb	53 W/kg / 31 lb/hp

Bajaj Platina Naked 100 cc 2006	6 kW / 8 bhp	113 kg / 249 lb	53 W/kg / 31 lb/hp
Subaru R2 type S 2003	47 kW / 63 bhp	830 kg / 1,830 lb	57 W/kg / 29 lb/hp
Ford Fiesta ECOnetic 1.6 L TDCi 5dr 2009	66 kW / 89 bhp	1,155 kg / 2,546 lb	57 W/kg / 29 lb/hp
Volvo C30 1.6D DRIVe S/S 3dr Hatch 2010	80 kW / 108 bhp	1,347 kg / 2,970 lb	59.4 W/kg / 27.5 lb/hp
Ford Focus ECOnetic 1.6 L TDCi 5dr Hatch 2009	81 kW / 108 bhp	1,357 kg / 2,992 lb	59.7 W/kg / 27 lb/hp
Ford Focus 1.8 L Zetec S TDCi 5dr Hatch 2009	84 kW / 113 bhp	1,370 kg / 3,020 lb	61 W/kg / 27 lb/hp
Honda FCX Clarity 4 kg Hydrogen 2008	100 kW / 134 bhp	1,600 kg / 3,528 lb	63 W/kg / 26 lb/hp
Hummer H1 6.6 L V8 2006	224 kW / 300 bhp	3,559 kg / 7,847 lb	63 W/kg / 26 lb/hp
Audi A2 1.4 L TDI 90 type S 2003	66 kW / 89 bhp	1,030 kg / 2,270 lb	64 W/kg / 25 lb/hp
Opel/Vauxhall/Holden/Chevrolet Astra 1.7 L CTDi 125 2010	92 kW / 123 bhp	1,393 kg / 3,071 lb	66 W/kg / 24.9 lb/hp
Mini (new) Cooper 1.6D 2007	81 kW / 108 bhp	1,185 kg / 2,612 lb	68 W/kg / 24 lb/hp
Toyota Prius 1.8 L 2010 (electric boost)	100 kW / 134 bhp	1,380 kg / 3,042 lb	72 W/kg / 23 lb/hp
Ford Focus 2.0 L Zetec S TDCi 5dr Hatch 2009	100 kW / 134 bhp	1,370 kg / 3,020 lb	73 W/kg / 23 lb/hp
General Motors EV1 electric car Gen II 1998	102.2 kW / 137 bhp	1,400 kg / 3,086 lb	73 W/kg / 23 lb/hp
Toyota Venza I4 2.7 L FWD 2009	136 kW / 182 bhp	1,706 kg / 3,760 lb	80 W/kg / 20.7 lb/hp
Ford Focus 2.0 L Zetec S 5dr Hatch 2009	107 kW / 143 bhp	1,327 kg / 2,926 lb	81 W/kg / 20 lb/hp
Fiat Grande Punto 1.6 L Multijet 120 2005	88 kW / 118 bhp	1,075 kg / 2,370 lb	82 W/kg / 20 lb/hp
Mini (classic) 1275GT 1969	57 kW / 76 bhp	686 kg / 1,512 lb	83 W/kg / 20 lb/hp
Opel/Vauxhall/Holden/Chevrolet Astra 2.0 L CTDi 160 2010	118 kW / 158 bhp	1,393 kg / 3,071 lb	85 W/kg / 19.4 lb/hp
Ford Focus 2.0 auto 2007	104.4 kW / 140 bhp	1,198 kg / 2,641 lb	87.1 W/kg / 19 lb/hp
Subaru Legacy/Liberty 2.0R 2005	121 kW / 162 bhp	1,370 kg / 3,020 lb	88 W/kg / 19 lb/hp
Subaru Outback 2.5i 2008	130.5 kW / 175 bhp	1,430 kg / 3,153 lb	91 W/kg / 18 lb/hp
Smart Fortwo 1.0 L Brabus 2009	72 kW / 97 bhp	780 kg / 1,720 lb	92 W/kg / 18 lb/hp
Toyota Venza V6 3.5 L AWD 2009	200 kW / 268 bhp	1,835 kg / 4,045 lb	109 W/kg / 15 lb/hp

Toyota Venza I4 2.7 L FWD 2009 with Lotus mass reduction	136 kW / 182 bhp	1,210 kg / 2,667 lb	112.2 W/kg / 14.7 lb/hp
Toyota Hilux V6 DOHC 4 L 4×2 Single Cab Pickup ute 2009	175 kW / 235 bhp	1,555 kg / 3,428 lb	112.5 W/kg / 14.6 lb/hp
Toyota Venza V6 3.5 L FWD 2009	200 kW / 268 bhp	1,755 kg / 3,870 lb	114 W/kg / 14.4 lb/ hp

Performance Luxury, Roadsters and Mild Sports, (Listed as Weight to Power)

Increased engine performance is a consideration, but also other features associated with luxury vehicles. Longitudinal engines are common. Bodies vary from hot hatches, sedans (saloons), coupés, convertibles and roadsters. Mid-range dual-sport and cruiser motorcycles tend to have similar power-to-weight ratios.

Vehicle	Power	Vehicle Weight	Weight-to-power ratio
Honda Accord sedan V6 2011	202 kW / 271 bhp	1630 kg / 3593 lb	124 W/kg / 13.26 lb/hp
Mini (new) Cooper 1.6T S JCW 2008	155 kW / 208 bhp	1205 kg / 2657 lb	129 W/kg / 13 lb/ hp
Mazda RX-8 1.3 L Wankel 2003	173 kW / 232 bhp	1309 kg / 2888 lb	141 W/kg / 12 lb/ hp
Holden Statesman/Caprice / Buick Park Avenue / Daewoo Veritas 6 L V8 2007	270 kW / 362 bhp	1891 kg / 4170 lb	143 W/kg / 12 lb/ hp
Kawasaki KLR650 Gasoline DualSport 650 cc	26 kW / 35 bhp	182 kg / 401 lb	143 W/kg / 11 lb/ hp
NATO HTC M1030M1 Diesel/Jet fuel DualSport 670 cc	26 kW / 35 bhp	182 kg / 401 lb	143 W/kg / 11 lb/ hp
Harley-Davidson FLSTF Softail Fat Boy Cruiser 1,584 cc 2009	47 kW / 63 bhp	324 kg / 714 lb	145 W/kg / 11.3 lb/ hp
BMW 7 Series 760Li 6 L V12 2006	327 kW / 439 bhp	2250 kg / 4960 lb	145 W/kg / 11 lb/hp
Subaru Impreza WRX STi 2.0 L 2008	227 kW / 304 bhp	1530 kg / 3373 lb	148 W/kg / 11 lb/ hp
Honda S2000 roadster 1999	183.88 kW / 240 bhp	1250 kg / 2723 lb	150 W/kg / 11 lb/ hp
GMH HSV Clubsport / GMV VXR8 / GMC CSV CR8 / Pontiac G8 6 L V8 2006	317 kW / 425 bhp	1831 kg / 4037 lb	173 W/kg / 9.5 lb/ hp
Tesla Roadster 2011	215 kW / 288 bhp	1235 kg / 2723 lb	174 W/kg / 9.5 lb/ hp

Sports Vehicles and Aircraft, (Listed as Weight to Power)

Power-to-weight ratio is an important vehicle characteristic that affects the acceleration and handling - and therefore the driving enjoyment - of any sports vehicle. Aircraft also depend on high power-to-weight ratio to achieve sufficient lift.

Vehicle	Power	Vehicle Weight	Weight-to-power ratio
Lotus Elise SC 2008	163 kW / 218 bhp	910 kg / 2006 lb	179 W/kg / 9 lb/hp
Ferrari Testarossa 1984	291 kW / 390 bhp	1506 kg / 3320 lb	193 W/kg / 9 lb/hp
Artega GT	220 kW / 300 bhp	1100 kg / 2425 lb	200 W/kg / 8 lb/hp
Lotus Exige GT3 2006	202.1 kW / 271 bhp	980 kg / 2160 lb	206 W/kg / 8 lb/hp
Chevrolet Corvette C6	321 kW / 430 bhp	1441 kg / 3177 lb	223 W/kg / 7 lb/hp
Suzuki V-Strom 650 V-twin DualSport 650 cc	50 kW / 67 bhp	194 kg / 427 lb	258 W/kg / 6.4 lb/hp
Chevrolet Corvette C6 Z06	376 kW / 505 bhp	1421 kg / 3133 lb	265 W/kg / 6.2 lb/hp
Porsche 911 GT2 2007	390 kW / 523 bhp	1440 kg / 3200 lb	271 W/kg / 6.1 lb/hp
Lamborghini Murciélago LP 670-4 SV 2009	493 kW / 661 bhp	1550 kg / 3417 lb	318 W/kg / 5.1 lb/hp
McLaren F1 GT 1997	467.6 kW / 627 bhp	1220 kg / 2690 lb	403 W/kg / 4.3 lb/hp
Bombardier Dash 8 Q400 turboprop airliner	7,562 kW / 10,142 bhp	17,185 kg / 37,888 lb	440 W/kg / 3.7 lb/hp
Supermarine Spitfire Fighter aircraft 1936	1,096 kW / 1,470 bhp	2,309 kg / 5,090 lb	475 W/kg / 3.46 lb/hp
Messerschmitt Bf 109 Fighter aircraft 1935	1,085 kW / 1,455 bhp	2,247 kg / 4,954 lb	483 W/kg / 3.40 lb/hp
Thunderbolt Land speed record car	3504 kW / 4700 bhp	7 t / 15432 lb	500 W/kg / 3.28 lb/hp
Ferrari FXX 2005	597 kW / 801 bhp	1155 kg / 2546 lb	517 W/kg / 3.2 lb/hp
Polaris Industries Assault Snowmobile 2009	115 kW / 154 bhp	221 kg / 487 lb	523 W/kg / 3.16 lb/hp
Ultima GTR 720 2006	536.9 kW / 720 bhp	920 kg / 2183 lb	583 W/kg / 3 lb/hp
Honda CBR1000RR 2009	133 kW / 178 bhp	199 kg / 439 lb	668 W/kg / 2.5 lb/hp
Ariel Atom 500 V8 2011	372 kW / 500 bhp	550 kg / 1212 lb	676.3 W/kg / 2.45 lb/hp
BMW S1000RR 2009	144 kW / 193 bhp	207.7 kg / 458 lb	693.3 W/kg / 2.37 lb/hp
Peugeot 208 T16 Pikes Peak 2013	652 kW / 875 bhp	875 kg / 1930 lb	745 W/kg / 2.2 lb/hp
KillaCycle Drag racing electric motorcycle	260 kW / 350 bhp	281 kg / 619 lb	925 W/kg / 1.77 lb/hp
MTT Turbine Superbike 2008	213.3 kW / 286 bhp	227 kg / 500 lb	940 W/kg / 1.75 lb/hp
Vyrus 987 C3 4V V supercharged motorcycle 2010	157.3 kW / 211 bhp	158 kg / 348.3 lb	996 W/kg / 1.65 lb/hp
BMW Williams FW27 Formula One 2005	690 kW / 925 bhp	600 kg / 1323 lb	1150 W/kg / 1.43 lb/hp
Honda RC211V MotoGP 2004-6	176.73 kW / 237 bhp	148 kg / 326 lb	1194 W/kg / 1.37 lb/hp

Boeing 747-300[dead link] at Mach 0.84 cruise, 35,000 ft altitude	245 MW / 328,656 bhp	178.1 t / 392,800 lb	1376 W/kg / 1.20 lb/hp
John Force Racing Funny Car NHRA Drag Racing 2008	5,963.60 kW / 8,000 bhp	1043 kg / 2,300 lb	5717 W/kg / 0.30 lb/hp
Koenigsegg One:1 2015	1000 kW / 1341 bhp	1310 kg / 2888 lb	763.35 W/kg / 2.15 lb/hp

Revolutions Per Minute

Revolutions per minute (abbreviated rpm, RPM, rev/min, r/min) is a measure of the frequency of rotation, specifically the number of rotations around a fixed axis in one minute. It is used as a measure of rotational speed of a mechanical component. In the French language, tr/mn (tours par minute) is the common abbreviation. The German language uses the abbreviation U/min or u/min (Umdrehungen pro Minute).

International System of Units

According to the International System of Units (SI), rpm is not a unit. This is because the word revolution is a semantic annotation rather than a unit. The annotation is instead done as a subscript of the formula sign if needed. Because of the measured physical quantity, the formula sign has to be f for (rotational) frequency and ω or Ω for angular velocity. The corresponding basic SI derived unit is s^{-1} or Hz. When measuring angular speed, the unit radians per second is used.

$$1\,\text{rad/s} \leftrightarrow \frac{1}{2\pi}\,\text{Hz}$$

$$\leftrightarrow \frac{60}{2\pi}\,\text{rpm}$$

$$1\,\text{rpm} \leftrightarrow \frac{1}{60}\,\text{Hz}$$

$$\leftrightarrow \frac{2\pi}{60}\,\text{rad/s}$$

$$1\,\text{Hz} \leftrightarrow 2\pi\,\text{rad/s}$$

$$\leftrightarrow 60\,\text{rpm}$$

Here the sign ↔ (correspondent) is used instead of = (equal). Formally, hertz (Hz) and radian per second (rad/s) are two different names for the same SI unit, s^{-1}. However, they are used for two different but proportional ISQ quantities: frequency and angular frequency (angular speed, magnitude of angular velocity). The conversion between a frequency f (measured in hertz) and an angular velocity ω (measured in radians per second) are:

$$\omega = 2\pi f\,,\ f = \frac{\omega}{2\pi}.$$

Thus a disc rotating at 60 rpm is said to be rotating at either 2π rad/s or 1 Hz, where the former measures the angular velocity and the latter reflects the number of revolutions per second.

If the non-SI unit rpm is considered a unit of frequency, then $1\,\text{rpm} = \frac{1}{60}\,\text{Hz}$. If it instead is considered a unit of angular velocity and the word "revolution" is considered to mean 2π radians, then $1\,\text{rpm} = \frac{2\pi}{60}\,\text{rad/s}$.

Examples

- On many kinds of disc recording media, the rotational speed of the medium under the read head is a standard given in rpm. Gramophone (phonograph) records, for example, typically rotate steadily at $16\,^2/_3$, $33\,^1/_3$, 45 or 78 rpm ($^5/_{18}$, $^5/_9$, $^3/_4$, or 1.3 Hz respectively).
- Modern ultrasonic dental drills can rotate at up to 800,000 rpm (13.3 kHz).
- The "second" hand of a conventional analogue clock rotates at 1 rpm.
- Audio CD players read their discs at a precise, constant rate (4.3218 Mbit/s of raw physical data for 1.4112 Mbit/s (176.4 kB/s) of usable audio data) and thus must vary the disc's rotational speed from 8 Hz (480 rpm) when reading at the innermost edge, to 3.5 Hz (210 rpm) at the outer edge.
- DVD players also usually read discs at a constant linear rate. The disc's rotational speed varies from 25.5 Hz (1530 rpm) when reading at the innermost edge, to 10.5 Hz (630 rpm) at the outer edge.
- A washing machine's drum may rotate at 500 to 2000 rpm (8–33 Hz) during the spin cycles.
- A power generation turbine (with a 2 pole alternator) rotates at 3000 rpm (50 Hz) or 3600 rpm (60 Hz), depending on country.
- Modern automobile engines are typically operated around 2000–3000 rpm (33–50 Hz) when cruising, with a minimum (idle) speed around 750–900 rpm (12.5–15 Hz), and an upper limit anywhere from 4500 to 10,000 rpm (75–166 Hz) for a road car or nearly 20,000 rpm for racing engines such as those in Formula 1 cars (currently limited to 15,000 rpm). The exhaust note of V8 F1 cars have a much higher pitch than an I4 engine, because each of the cylinders of a four-stroke engine fires once for every two revolutions of the crankshaft. Thus an eight-cylinder engine turning 300 times per second will have an exhaust note of 1200 Hz.
- A piston aircraft engine typically rotates at a rate between 2000 and 3000 rpm (30–50 Hz).
- Computer hard drives typically rotate at 5400 or 7200 rpm (90 or 120 Hz), the most common speeds for the ATA or SATA-based drives in consumer models. High-performance drives (used in fileservers and enthusiast-gaming PCs) rotate at 10,000 or 15,000 rpm (160 or 250 Hz), usually with higher-level SATA, SCSI or Fibre Channel interfaces and smaller

platters to allow these higher speeds, the reduction in storage capacity and ultimate outer-edge speed paying off in much quicker access time and average transfer speed thanks to the high spin rate. Until recently, lower-end and power-efficient laptop drives could be found with 4200 or even 3600 rpm spindle speeds (70 and 60 Hz), but these have fallen out of favour due to their lower performance, improvements in energy efficiency in faster models and the takeup of solid-state drives for use in slimline and ultraportable laptops. Similar to CD and DVD media, the amount of data that can be stored or read for each turn of the disc is greater at the outer edge than near the spindle; however, hard drives keep a constant rotational speed so the effective data rate is faster at the edge (conventionally, the "start" of the disc, opposite to CD/DVD).

- Floppy disc drives typically ran at a constant 300 or occasionally 360 rpm (a relatively slow 5 or 6 Hz) with a constant per-revolution data density, which was simple and inexpensive to implement, though inefficient. Some designs such as those used with older Apple computers (Lisa, early Macintosh, later II's) were more complex and used variable rotational speeds and per-track storage density (at a constant read/record rate) to store more data per disc; for example, between 394 rpm (with 12 sectors per track) and 590 rpm (8 sectors) with the Mac's 800 KB double-density drive at a constant 39.4 KB/s (max) – vs. 300 rpm, 720 KB and 23 KB/s (max) for double-density drives in other machines.
- A Zippe-type centrifuge for enriching uranium spins at 90,000 rpm (1,500 Hz) or faster.
- Gas turbine engines rotate at tens of thousands of rpm. JetCat model aircraft turbines are capable of over 100,000 rpm (1,700 Hz) with the fastest reaching 165,000 rpm (2,750 Hz).
- A Flywheel energy storage system works at 60,000–200,000 rpm (1–3 kHz) range using a passively magnetic levitated flywheel in vacuum. The choice of the flywheel material is not the most dense, but the one that pulverises the most safely, at surface speeds about 7 times the speed of sound.
- A typical 80 mm, 30 CFM computer fan will spin at 2,600–3,000 rpm on 12 V DC power.
- A millisecond pulsar can have near 50,000 rpm.
- A turbocharger can reach 290,000 rpm (4.8 kHz), while 80,000–200,000 rpm (1–3 kHz) is common.
- Molecular microbiology - molecular engines. The rotation rates of bacterial flagella have been measured to be 10,200 rpm (170 Hz) for Salmonella typhimurium, 16,200 rpm (270 Hz) for Escherichia coli, and up to 102,000 rpm (1,700 Hz) for polar flagellum of Vibrio alginolyticus, allowing the latter organism to move in simulated natural conditions at a maximum speed of 540 mm per hour.

References

- Bornschlegl, Susanne (2012). Ready for SIL 4: Modular Computers for Safety-Critical Mobile Applications (pdf). MEN Mikro Elektronik. Retrieved 2015-09-21
- George Bibel. Beyond the Black Box: the Forensics of Airplane Crashes. Johns Hopkins University Press, 2008. ISBN 0-8018-8631-7
- Ofria, Charles. "A short course on automatic transmissions". CarParts.com. JC Whitney. Archived from the original on 6 October 2014. Retrieved 6 October 2014

- "NASA Physiological Acceleration Systems". Web.archive.org. 2008-05-20. Archived from the original on 2008-05-20. Retrieved 2012-12-25
- Snyder, John Beltz (2014-07-10). "2015 Nissan Leaf gets B mode standard, new MorningSky Blue color". AutoBlog. Retrieved 2016-03-21

Permissions

All chapters in this book are published with permission under the Creative Commons Attribution Share Alike License or equivalent. Every chapter published in this book has been scrutinized by our experts. Their significance has been extensively debated. The topics covered herein carry significant information for a comprehensive understanding. They may even be implemented as practical applications or may be referred to as a beginning point for further studies.

We would like to thank the editorial team for lending their expertise to make the book truly unique. They have played a crucial role in the development of this book. Without their invaluable contributions this book wouldn't have been possible. They have made vital efforts to compile up to date information on the varied aspects of this subject to make this book a valuable addition to the collection of many professionals and students.

This book was conceptualized with the vision of imparting up-to-date and integrated information in this field. To ensure the same, a matchless editorial board was set up. Every individual on the board went through rigorous rounds of assessment to prove their worth. After which they invested a large part of their time researching and compiling the most relevant data for our readers.

The editorial board has been involved in producing this book since its inception. They have spent rigorous hours researching and exploring the diverse topics which have resulted in the successful publishing of this book. They have passed on their knowledge of decades through this book. To expedite this challenging task, the publisher supported the team at every step. A small team of assistant editors was also appointed to further simplify the editing procedure and attain best results for the readers.

Apart from the editorial board, the designing team has also invested a significant amount of their time in understanding the subject and creating the most relevant covers. They scrutinized every image to scout for the most suitable representation of the subject and create an appropriate cover for the book.

The publishing team has been an ardent support to the editorial, designing and production team. Their endless efforts to recruit the best for this project, has resulted in the accomplishment of this book. They are a veteran in the field of academics and their pool of knowledge is as vast as their experience in printing. Their expertise and guidance has proved useful at every step. Their uncompromising quality standards have made this book an exceptional effort. Their encouragement from time to time has been an inspiration for everyone.

The publisher and the editorial board hope that this book will prove to be a valuable piece of knowledge for students, practitioners and scholars across the globe.

Index

A

Adiabatic Compression, 40
Adiabatic Expansion, 47
Aerodynamic Drag, 212
Air-fuel Mixture, 8-9, 21, 23, 51, 55-56, 59, 62, 170
Automatic Transmission, 2, 174-183, 185-190
Automotive Engines, 1-3, 167

C

Cylinder Liner, 4

D

Diesel Cycle Engine, 40, 50
Diesel Engines, 3, 29, 31, 33, 35, 40-41, 46, 50, 88, 95, 158, 215
Diesel Locomotive, 223

E

Engine Configuration, 70-71
Engine Control Unit, 81, 172-173, 180, 184
Engine Sensors, 79, 81-82
Engine Speed Sensor, 82, 85
Engine Valves, 4, 21
Engine Work, 37, 39
Enthalpy Flux, 39
Ethanol, 92, 98, 106-107, 157, 213
Exhaust Gas Recirculation, 170
Exhaust Stroke, 37, 42, 47-48, 52, 59-60, 67
Exhaust System, 9, 33, 53, 87
External Combustion Engines, 27, 77

F

Flywheel, 3, 5, 19-21, 33-34, 51-52, 66-67, 230
Flywheels, 4, 105
Four Stroke Diesel Engine, 46
Four-stroke Cycle, 37, 51, 53, 55
Fuel Filters, 4, 33-34
Fuel Injection, 33, 47, 172
Fuel Mixture, 3-4, 8-9, 21, 23, 27, 51, 53, 55-56, 59, 62, 81-83, 170

G

Gas Turbine Engines, 70, 100, 230
Gasoline Engines, 3, 32, 50, 52, 93, 158
Gasoline Motors, 3
Grey Cast Iron, 5, 7

H

Hall Effect Sensor, 86
Heavy Motor Vehicle, 1
Hybrid Vehicles, 181, 212, 217

I

Ideal Gas, 43-45, 119-120, 137, 147, 156-157
Ignition Distributor, 5, 20, 67
Ignition System, 172
Ignition Timing, 82, 170
Inlet Valve, 20, 27, 55, 59
Inline Engine, 70
Inline Engines, 68
Intake Manifold, 3, 9, 67, 81-82
Intake Stroke, 37, 47-48, 51, 56, 59
Internal Combustion Engine, 1, 4, 24, 27, 33-34, 37-38, 40, 51, 53, 56, 60-61, 65, 67, 70, 92-93, 175
Internal Combustion Engines, 25, 27, 29, 53, 92-93
Internal Energy, 42-45, 119, 129, 133-135, 137-144
Isentropic Compression, 42
Isentropic Expansion, 42
Isobaric Heat Addition, 42, 48
Isochoric Process, 41, 43-44, 46, 49, 136

J

Jet Engines, 100
Jurisdiction, 169

K

Kinetic Energy, 106, 142-143, 162-163, 209, 211-212

L

Lubrication System, 8, 33, 172

M

Magnetic Field, 86
Manual Transmission, 2, 175-177, 179, 188-191
Mating Surfaces, 5, 9
Motor Cooling System, 3

N
Natural Gas Engines, 35
Nitrogen Oxide, 95
Nitrous Oxide, 91

O
Oil Filters, 4, 33-34
Oil Galleries, 5
Otto Cycle, 35, 37-39, 41, 50, 58-59
Oxygen Sensor, 81, 87-88

P
Petrol Engine, 32, 75, 215
Piston Engines, 3
Piston Ring, 4, 15
Poppet Valve, 4, 21
Poppet Valves, 76
Power Stroke, 13, 20, 37, 40, 42, 47, 52, 55-56, 58-59, 67
Producer Gas Engines, 35, 37
Pushrods, 24

R
Radial Engine, 67-70, 223
Reciprocating Engine, 27, 29-31
Rod Bearings, 19
Rotary Engine, 31
Rotary Valve, 4, 21, 23
Rotating Crankshaft, 2

S
Self-ignition Engines, 3
Spark Ignition, 29-30, 40
Spark Plugs, 3, 6, 32, 40
Standby Generators, 31
Steam Engines, 77, 106, 128

T
Thermal Efficiency, 38-39, 41, 47-50, 56, 60, 66, 211
Thermal Output, 29-30
Throttle Position Sensor, 81
Throttle Valve, 173, 180
Timing Belt Cover, 25-26
Timing Gear, 5, 25
Top Dead Centre, 12, 28
Turbocharger, 34, 79, 172, 230

V
V-engines, 3
Variable Resistor, 81-82
Variable Transmissions, 181

W
W Engine, 73-75
Wankel Engine, 31, 223
Waste Feed, 36

www.ingramcontent.com/pod-product-compliance
Lightning Source LLC
LaVergne TN
LVHW080722210926
848485LV00003B/507